MATHEMATICAL
METHODS
FOR CAD

MATHEMATICAL METHODS FOR CAD

J.J. RISLER

Université Pierre et Marie Curie
École normale supérieure
France

CAMBRIDGE UNIVERSITY PRESS

Cambridge

New York   Port Chester

Melbourne   Sydney

Published by the Press Syndicate of the University of Cambridge
The Pitt Building, Trumpington Street, Cambridge CB2 1RP
40 West 20th Street, New York, NY 10011-4211, USA
10 Stamford Road, Oakleigh, Victoria 3166, Australia

Published with the support of
the Ministère de la Recherche et de la Technologie (France) :
« Programme d'aide à l'édition d'ouvrages scientifiques et techniques »

**This book appears in French in the series
"Recherches en Mathématiques Appliquées"
edited by P.G. Ciarlet and J.L. Lions (1991)**

Library of Congress cataloging in publication data available

A catalogue record for this book is available from the British Library

ISBN 0 521 43100X hardback
ISBN 0 521 436915 paperback

Printed in France

Contents

Preface

The content of this book was taught three years in the "D.E.A. d'Analyse Numérique de l'Université Pierre et Marie Curie" (Paris). The book appeared in French in 1990 (Méthodes Mathématiques pour la CAO, RMA 18, Masson). This book is an introduction to CAD from a mathematical and thoretical point of view. The assertions and results are all proved, and the practical considerations (such as implementation, programming language, description of up to date systems of CAD, etc ...) are not studied.

The most important part of the book is the study of "B-splines" functions, which are more and more used these days. Their remarkable mathematical properties are studied in detail, and also the geometric properties of the curves and surfaces one may define with them. "Bézier" curves, historically used before B-splines, are introduced here as special B-spline curves.

Following an idea of C. De Boor and Hölig ([DB-H]), B-splines are defined without the use of divided differences, which lightens the exposition. However, a section recalls the definition and the main properties of divided differences, in order that the reader could be able to refer easily to books adopting this viewpoint. Moreover, the definition of B-splines with divided differences may be generalized in a natural way to the several variables case, which permits the definition of polyhedral splines.

Surfaces are studied from the viewpoints "tensor product" and "triangular Bézier patches". "Polyhedral" spline surfaces (and in particular "box-splines") are also considered, even though they are not yet used in the existing CAD systems.

I have also included a little "algorithmic geometry", which is also a useful domain for CAD. The only question treated here in this regard, which is one of the main tools of the theory, is the problem of automatic triangulation (with a method known under the name of "Voronoï-Delaunay").

In the last part, some algebraic problems are considered, and a brief introduction to the properties of polynomials from the point of view of formal computation is given. This matter is at the moment the subject of much study, and in my opinion will soon become an essential tool for CAD, especially for the resolution of specific problems, such as the intersection of surfaces for instance, when classical methods are deficient. This part is only an introduction to the area, giving the occasion to go over again or to learn some fundamental algebraic techniques (such as Sturm sequences, or resultants) essential for formal computation.

The book is organized as follows.

Chapter 1 introduces non-uniform B-splines (i.e., defined with any knot sequence of $\mathbf{R}$) and the basic algorithms used in the theory (evaluation at a point, differentiation algorithm, insertion of a new knot). The "regularizing" property of B-splines is shown, and the chapter ends with a presentation of the "divided differences" point of view.

Chapter 2 is devoted to the study of B-splines and Bézier curves (these are introduced as a particular case of B-spline curves), and to the geometric signifiance of the algorithms given in Chapter 1. Several subdivision algorithms (such as the "Oslo" algorithm for instance) are given, in the general case of B-splines, as well as in the Bézier curve case.

Chapter 3 treats the interpolation problem of a family of points by a spline curve, and of several complements, such as geometric continuity, rational spline or Bézier curves, rational representations of conics, etc ...

Chapter 4 is devolved to surfaces, mainly from the classical point of view of bicubic splines (i.e., tensor products of cubic B-splines), and on the triangular Bézier patches. The rational point of view is also studied, especially the problem and the usefulness of base points. The chapter ends with a rather complete study of polyhedral splines and box-splines.

Chapter 5 presents some tools of algorithmic geometry, especially the problem of triangulation (hopefully automatic) of a domain (not necessarily plane), when the vertices of simplices are fixed. The method of "Voronoï-Delaunay" is explained; this method is known to be optimal (or at least a good one) from the point of view of the shape of the simplices, at least in the case of the plane.

Chapter 6 is the last one and gives an insight into the theory of formal computation, and its relations with geometric problems. In particular, some properties of roots of real polynomials are studied, as well as the resultant of two polynomials, and the notion of real semi-algebraic sets.

I thank the staff of the " Laboratoire d'Analyse Numérique de l'Université Paris 6" for their welcome, and its director Philippe Ciarlet who invited me to teach CAD in the frame of" D.E.A. d'Analyse Numérique".

I thank also Professors Fiorot and Sablonnière, who introduced me to the subject of B-splines in the Rennes Seminar of May 1987 (see [Re]). I used much ofthe material of this seminar in the early chapters of this book. I thank also Michel Merle who gave me his personal notes about polyhedral splines.

1

B-splines

1.1 PIECEWISE POLYNOMIAL FUNCTIONS

1.1.1 Notation   Let $\mathcal{P}_k$ denote the vector space (over the field $\mathbf{R}$) of polynomials of degree $\leq k$ in one variable. $\mathcal{P}_k$ is then of dimension $k+1$ over $\mathbf{R}$.

Let also be given an interval $[a, b] \subset \mathbf{R}$, an integer $\ell \geq 1$, a set τ of $\ell - 1$ points (τ_i) $(1 \leq i \leq \ell - 1)$ in $]a, b[$

$$a < \tau_1 < \cdots < \tau_{\ell-1} < b$$

called *knots*, and $\ell - 1$ integers r_i such that $0 \leq r_i \leq k$. We will set $a = \tau_0$, $b = \tau_\ell$.

$\mathcal{P}_{k,\tau,r}$ will denote the vector space of the piecewise polynomial functions of degree $\leq k$ on $[a, b]$, with $C^{r_i - 1}$ continuity at τ_i $(1 \leq i \leq \ell - 1)$ (this means that the function has $r_i - 1$ continuous derivatives at τ_i; for $r_i = 0$ there is no condition).

1.1.2 Lemma

$$dim \mathcal{P}_{k,\tau,r} \;\;=\;\; (k+1)\ell - \sum_{i=1}^{\ell-1} r_i$$

Proof   The space of functions, whose restriction to each of the ℓ intervals $[\tau_i, \tau_{i+1}]$ is a polynomial of degree k, has dimension $(k+1)\ell$.

A basis for this space is given by the functions f_{ij} such that $f_{ij} = (X - \tau_i)^j$ for $X \in [\tau_i, \tau_{i+1}]$ and $f_{ij} = 0$ for $x \notin [\tau_i, \tau_{i+1}]$ $(0 \le i \le \ell - 1, \quad 0 \le j \le k)$.

Let $f \in \mathcal{P}_{k,\tau,r}$, f be equal to the polynomial P_i on $[\tau_i, \tau_{i+1}]$, with $P_i = \sum_{j=0}^{k} a_j^i (X - \tau_i)^j$; the condition for f to be $\mathcal{C}^{r_i-1}$ at τ_i $(1 \le i \le \ell - 1)$ introduces r_i linear relations between the a_i^j's:

$$\begin{cases} a_o^i = \phi_o(a_o^{i-1}, \ldots, a_k^{i-1}) \\ \quad \vdots \\ a_{r_i-1}^i = \phi_{r_i-1}(a_o^{i-1}, \ldots, a_k^{i-1}). \end{cases}$$

These equations are clearly independent, as each one introduces a new variable.

∎

1.1.3 A basis for $\mathcal{P}_{k,\tau,r}$

Let $(X - \tau_i)_+$ be the function equal to $(X - \tau_i)$ for $X \ge \tau_i$ and 0 for $X \le 0$; one then has $(X - \tau_i)_+ = \mathrm{Sup}(X - \tau_i, 0)$.

1.1.4 Lemma *The set of restrictions to $[a, b]$ of the functions $(X - \tau_i)_+^j$ $(0 \le i \le \ell - 1,\ r_i \le j \le k)$ is a basis of $\mathcal{P}_{k,\tau,r}$.*

Proof Since $(X - \tau_i)_+$ is of class $\mathcal{C}^o$, it is clear by induction on j that $(X - \tau_i)_+^j$ is of class $\mathcal{C}^{j-1}$ at τ_i, so that $(X - \tau_i)_+^j \in \mathcal{P}_{k,\tau,r}$ for $r_i \le j \le k$; as there are $(k+1)\ell - \sum_{i=1}^{\ell-1} r_i$ such functions, it is enough to see that the $(X - \tau_i)_+^j$'s are linearly independent, which is clear.

∎

1.2 AN EXAMPLE: CUBIC SPLINES

Let $\mathcal{S}_{k,\tau}$ denote the space $\mathcal{P}_{k,\tau,r}$ where $r_i = k$ $(1 \le i \le \ell - 1)$; traditionally $\mathcal{S}_{k,\tau}$ is called the space of splines of degree k with knots τ_i; one then has

$$\dim \mathcal{S}_{k,\tau} = \ell(k+1) - \sum_{i=1}^{\ell-1} k = k + \ell,$$

and the splines are $\mathcal{C}^{k-1}$ functions.

The most usual case is that of cubic splines.

1.2.1 Proposition   Let $M_i = (\tau_i, y_i)$ $(0 \leq i \leq \ell)$ be $\ell + 1$ points in $\mathbf{R}^2$; if two numbers α and β are given, there exists a unique spline function $f \in S_{3,\tau}$ such that the curve $y = f(x)$ passes through the points M_i and satisfies

$$\begin{cases} f'(a) = \alpha \\ f'(b) = \beta. \end{cases}$$

Proof   This proposition is introductory so we will give the proof only in the (easier) case of a uniform spacing of the knots τ_i $(\tau_{i+1} - \tau_i = h)$. The above conditions give $\ell + 3$ linear conditions

$$\begin{cases} f(\tau_i) = y_i & (0 \leq i \leq \ell) \\ f'(a) = \alpha \\ f'(b) = \beta. \end{cases}$$

The dimension of $S_{3,\tau}$ is $\ell + 3$, so it is enough to see that these conditions are linearly independent.

For that, it is convenient to write the restriction P_i of f to $[\tau_i, \tau_{i+1}]$ in the following way

$$P_i(u) = a_i + b_i u + c_i u^2 + d_i u^3 \quad (0 \leq i \leq \ell - 1)$$

with $0 \leq u \leq 1$, $u = \frac{t - \tau_i}{h}$, and

$$\begin{cases} P_i(0) = f(\tau_i) = y_i \\ P_i(1) = f(\tau_{i+1}) = y_{i+1} & (0 \leq i \leq \ell - 1). \end{cases}$$

The conditions on a_i, b_i, c_i, d_i are then

(1) $$a_i = y_i$$

(2) $$a_i + b_i + c_i + d_i = y_{i+1} \quad (0 \leq i \leq \ell - 1)$$

(with $f(\tau_i) = y_i$)

(3) $$\begin{cases} b_o = \alpha/h \\ b_{\ell-1} + 2c_{\ell-1} + 3d_{\ell-1} = \beta/h \end{cases}$$

(boundary conditions), and the fact that the function is C^2

(4) $$b_i + 2c_i + 3d_i = b_{i+1}$$

(5) $$2c_i + 6d_i = 2c_{i+1} \quad (1 \le i \le \ell - 2)$$

which gives immediately by (1), (2) and (4)

$$\begin{cases} c_i = 3(y_{i+1} - y_i) - 2b_i - b_{i+1} \\ d_i = -2(y_{i+1} - y_i) + b_i + b_{i+1} \end{cases}$$

and upon inserting these formulas in (5) (for index $i + 1$),

$$b_{i-1} + 4b_i + b_{i+1} = 3(y_{i+1} - y_{i-1}) \qquad (1 \le i \le \ell - 1)$$

which gives the following system to compute the b_i's

$$
\begin{pmatrix}
1 & & & & \\
1 & 4 & 1 & & \\
 & 1 & 4 & 1 & \\
 & & & \ddots & \\
 & & & & \ddots \\
 & & & 1 & 4
\end{pmatrix}
\begin{pmatrix}
b_o \\ \cdot \\ \cdot \\ \cdot \\ \cdot \\ b_{\ell-1}
\end{pmatrix}
=
\begin{pmatrix}
\alpha/h \\ 3(y_2 - y_o) \\ 3(y_3 - y_1) \\ \cdot \\ \cdot \\ 3(y_\ell - y_{\ell-2}) - \beta/h
\end{pmatrix}.
$$

The matrix of this system is clearly nonsingular, as is easily seen by induction (one may also apply Theorem 3.1.1). The fact that the principal minors are all non zero implies that one may apply the Gauss method without pivoting, i.e., that at the ith step, the ith element of the ith row is non zero and so may be taken as the ith pivot.

One sees also that the solution of this system is very fast: the number of operations required is linear in ℓ.

The use of the word spline comes from the following theorem.

1.2.2 Theorem   Let $f \in S_{3,\tau}$ be the spline function such that

$$\begin{cases} f(\tau_i) = y_i & (0 \le i \le \ell) \\ f'(a) = \alpha \\ f'(b) = \beta & \text{(see 1.2.1).} \end{cases}$$

Then if E denotes the set of C^r functions on $[a, b]$ such that

$$
\begin{cases}
\phi(\tau_i) = y_i & (0 \le i \le \ell) \\
\phi'(a) = \alpha \\
\phi'(b) = \beta
\end{cases}
$$

f is the only element of E which minimizes (among the elements of E) the integral $\int_a^b [\phi'']^2(t) \, dt$.

Proof Let $\phi \in E$; set $e = \phi - f$ ("error in the approximation of ϕ by the spline function f").

Let $S_{1,\tau}$ be the space of the "splines of degree one on $[a, b]$ with knots at τ_i", i.e., the C° functions, piecewise linear, with change of slope at the points τ_i.

1.2.3 Lemma

$$
\int_a^b e''(x) \, h(x) \, dx = 0 \quad \forall\, h \in S_{1,\tau}
$$

Proof

$$
\int_a^b e''(x) \, h(x) \, dx = \sum_{i=0}^{\ell-1} \int_{\tau_i}^{\tau_{i+1}} e''(x) \, h(x) \, dx
$$

$$
= \sum_{i=0}^{\ell-1} \left((e'h)_{\tau_i}^{\tau_{i+1}} - \int_{\tau_i}^{\tau_{i+1}} e'h' \right)
$$

(integration by parts). But we have

$$
\sum_{i=0}^{\ell-1} (e'h)_{\tau_i}^{\tau_{i+1}} = e'h(b) - e'h(a) = 0
$$

because $e'(a) = e'(b) = 0$ by the definition of E, and

$$
\sum_{i=0}^{\ell-1} \int_{\tau_i}^{\tau_{i+1}} e'h' = \sum_{i=0}^{\ell-1} \lambda_i (e(\tau_{i+1}) - e(\tau_i)) \quad \text{if } h'(x) = \lambda_i \quad \text{for } x \in]\tau_i, \tau_{i+1}[,
$$

and we have $e(\tau_i) = 0$, as by definition, $f(\tau_i) = \phi(\tau_i) \quad (0 \le i \le \ell)$.

∎

Completion of the proof of Theorem 1.2.2

Let $\phi \in E$, $e = \phi - f$; we have

$$\int_a^b \phi''^2 = \int_a^b (e'' + f'')^2 = \int_a^b e''^2 + \int_a^b f''^2 + 2 \int_a^b e'' f''.$$

But since f is in $S_{3,\tau}$, one has $f'' \in S_{1,\tau}$, and then $\int_a^b e'' f'' = 0$ by Lemma 1.2.3; one thus has $\int_a^b \phi''^2 \geq \int_a^b f''^2$, and if there is equality, one deduces $\phi'' = f''$ (as then $\int_a^b e''^2 = 0$ and so $e'' = 0$ because e'' is continuous), which implies immediately $\phi = f$ if $\ell \geq 1$, because ϕ and f satisfy $\phi'(\alpha) = f'(\alpha)$, $\phi'(\beta) = f'(\beta)$ and $\phi(\tau_i) = f(\tau_i)$.

■

1.3 B-SPLINES: FUNDAMENTAL PROPERTIES

We now generalize the spline functions of Section 1.2, by considering the space $\mathcal{P}_{k,\tau,r}$ (see 1.1). In other words,

a) k will take any value (i.e., not necessarily 3),

b) the functions will be of class C^{r_i-1} at the point τ_i ($1 \leq i \leq \ell - 1$) with $r_i \leq k$ (and not necessarily $r_i = k$).

To be able to describe a manageable base of this space, we shall first define B-spline functions.

1.3.1 Notation   Let us take in **R** a sequence $t_0, \ldots, t_m$ of points called *knots* such that for all i, $t_i \leq t_{i+1}$. If there are r t_i's equal to τ, one says that τ is a node *of order r* or *of multiplicity r*.

Moreover, for $1 \leq j \leq m + 1 - i$, one sets:

$$\omega_{i,j}(x) = \begin{cases} \dfrac{x - t_i}{t_{i+j} - t_i} & \text{if } t_i < t_{i+j} \\ 0 & \text{otherwise.} \end{cases}$$

1.3.2 Definition of B-splines

With the above notation, set $t = (t_0, \ldots, t_m)$. For $x \in \mathbf{R}$, $0 \leq i \leq m - k - 1$, the functions $B_{i,k,t}(x)$ (also denoted $B_{i,k}(x)$ when *the sequence t is fixed) are defined by induction on k in the following way*

$$
\begin{cases}
B_{i,0}(x) = \begin{cases} 1 & \text{if } t_i \leq x < t_{i+1} \\ 0 & \text{otherwise} \end{cases} \\
\\
B_{i,k}(x) = \omega_{i,k}(x)B_{i,k-1}(x) + \left(1 - \omega_{i+1,k}(x)\right)B_{i+1,k-1}(x) \text{ for } k \geq 1.
\end{cases}
$$

By definition of $\omega_{i,k}(x)$, one has

$$
B_{i,k}(x) = \frac{x - t_i}{t_{i+k} - t_i} B_{i,k-1}(x) + \frac{t_{i+k+1} - x}{t_{i+k+1} - t_{i+1}} B_{i+1,k-1}(x)
$$

if $t_i < t_{i+k}$ and $t_{i+1} < t_{i+k+1}$.

1.3.3 Remarks

1) One may define in the same way B-splines for an *infinite* sequence of knots t_i $(t_i \leq t_{i+1})$, since the definition of each $B_{i,k}$ uses only a finite numbers of knots (see Proposition 1.3.4 below: the spline $B_{i,k}(x)$ has $[t_i, t_{i+k+1}]$ for support, and its definition uses only those t_j such that $i \leq j \leq i + k + 1$).

2) If, for an index i, $t_i = t_{i+k+1}$ (and so t_i is a node of multiplicity $\geq k + 2$), one has $B_{i,k} \equiv 0$; the converse is also true: see 1.3.4 below.

1.3.4 Proposition   *With the above notation, one has the following properties:*

a) $B_{i,k}(x)$ *is a piecewise polynomial of degree k.*

b) $B_{i,k}(x) = 0$ *for* $x \notin [t_i, t_{i+k+1}[$.

c) $B_{i,k}(x) > 0$ *for* $x \in]t_i, t_{i+k+1}[$; $B_{i,k}(t_i) = 0$ *unless* $t_i = t_{i+1} = \cdots = t_{i+k} < t_{i+k+1}$ *(node of order $k + 1$), and then* $B_{i,k}(t_i) = 1$.

d) *Let* $[a, b]$ *be an interval such that* $t_k \leq a$, $t_{m-k} \geq b$. *Then* $\sum_{i=0}^{m-k-1} B_{i,k}(x) = 1$ *for all* $x \in [a, b[$.

e) *Let* $x \in]t_i, t_{i+k+1}[$. *Then* $B_{i,k,t}(x) = 1$ *if and only if* $t_{i+1} = \cdots = t_{i+k} = x$.

f) $B_{i,k}(x)$ *is right-continuous (and even right-infinitely differentiable), for all* $x \in \mathbf{R}$ *(recall that* $B_{i,k}(x) = 0$ *for x outside* $[t_0, t_m]$*).*

1.3.5 Remark   One sees already in this proposition some remarkable properties of the B-splines: d) shows that they constitute a *partition of*

unity on a convenient interval, and b) shows that each $B_{i,k}(x)$ has a "*small support*".

Proof Properties a), b), c), d) and f) are clear for $k = 0$; one deduces immediately a), b), c) and f) for $B_{i,k}$ by induction on k.

Let us prove property d) by induction on k. Let $x \in [a, b]$. There exists then j, $j \geq k$, $j \leq m - k - 1$, such that $x \in [t_j, t_{j+1}[$.

If $x = t_j$ and $B_{j,k}(x) = 1$, the property is clear (see c)). In the other cases, one has

$$\sum_{i=0}^{m-k-1} B_{i,k}(x) = \sum_{i=j-k}^{j} B_{i,k}(x)$$

by b), and

$$\sum_{j-k}^{j} B_{i,k}(x) = \sum_{j-k}^{j} \omega_{i,k} B_{i,k-1}(x) + \sum_{j-k}^{j} (1 - \omega_{i+1,k}) B_{i+1,k-1}(x)$$

by the definition of the B-splines, which implies, grouping terms together,

$$\sum_{i=j-k}^{j} B_{i,k}(x) = \omega_{j-k} B_{j-k,k-1}(x)$$

$$+ \sum_{i=j+1-k}^{j} B_{i,k-1}(x) + (1 - \omega_{j+1,k}) B_{j+1,k-1}(x).$$

But $B_{j-k,k-1}(x) = 0$ and $B_{j+1,k-1}(x) = 0$ because $x \in [t_j, t_{j+1}[$ (see b)) and

$$\sum_{i=j+1-k}^{j} B_{i,k-1}(x) = 1$$

by the induction hypothesis, which implies d).

e) is an easy consequence of d) and c).

∎

1.3.6 Remark If $t_{m-k} = \cdots = t_m = b$, formula d) is valid only on the interval $[a, b[$, because for all i ($1 \leq i \leq m - k - 1$), one has $B_{i,k}(b) = 0$ (see f) above: B-splines are right-continuous). If we want a formula valid on $[a, b]$, we have to set $B_{m-k-1,k}(b) = 1$, which makes the B-spline $B_{m-k-1,k}(x)$ *left continuous* at b.

We will systematically use this convention, which is necessary only if there is a knot of multiplicity $k + 1$ at b.

1.3.7 Examples

a) $k = 1$

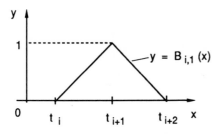 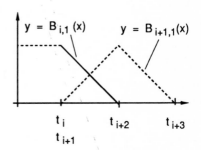

Case of simple knots   Case of a double knot

Figure 1.3.1

b) $k = 2$ (i.e., piecewise quadratic functions)

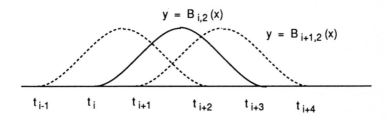

Case of simple knots

Figure 1.3.2

(The curve $y = B_{i,2}(x)$ is built from three arcs of parabolas and two half-lines with junction C^1 at the knots t_i, t_{i+1}, t_{i+2} and t_{i+3}).

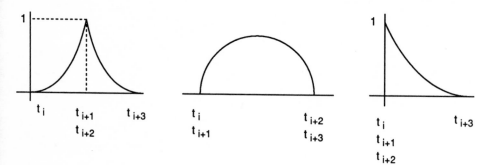

$y = B_{i,2}(x)$ in the case of multiple knots

Figure 1.3.3

c) Let us represent the 9 elements of the B-spline basis of $\mathcal{P}_{3,t}$ when $n = 9$, i.e., with 5 knots (simple) in $]a, b[$, and $t_0 = \cdots = t_3 = a$ and $t_9 = \cdots = t_{12} = b$

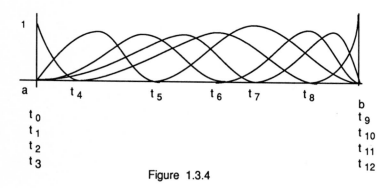

Figure 1.3.4

1.3.8 Remark We will give later (Chap. 3, Section 3.3) the expressions for the B-splines $B_{i,3}(x)$ in the case of an uniform sequence of knots .

1.3.9 Proposition   *For all $k \geq 0$ and all $x \in \mathbf{R}$, $B_{i,k}$ is right-differentiable and one has*

$$B'_{i,k}(x) \;=\; k\left[\frac{B_{i,k-1}(x)}{t_{i+k}-t_i} \;-\; \frac{B_{i+1,k-1}(x)}{t_{i+k+1}-t_{i+1}}\right]$$

where, by convention, one replaces by 0 an expression whose denominator is equal to zero.

Proof   By induction on k using Definition 1.3.2, the case $k = 0$ being trivial.

We have by definition:

$$B_{i,k} \;=\; \frac{x-t_i}{t_{i+k}-t_i}B_{i,k-1} \;+\; \frac{t_{i+k+1}-x}{t_{i+k+1}-t_{i+1}}B_{i+1,k-1}$$

and, differentiating and using induction:

$$B'_{i,k} = \frac{B_{i,k-1}}{t_{i+k}-t_i} - \frac{B_{i+1,k-1}}{t_{i+k+1}-t_{i+1}}$$
$$+(k-1)\left(\frac{x-t_i}{t_{i+k}-t_i}\left[\frac{B_{i,k-2}}{t_{i+k-1}-t_i} - \frac{B_{i+1,k-2}}{t_{i+k}-t_{i+1}}\right]\right.$$
$$\left.+\frac{t_{i+k+1}-x}{t_{i+k+1}-t_{i+1}}\left[\frac{B_{i+1,k-2}}{t_{i+k}-t_{i+1}} - \frac{B_{i+2,k-2}}{t_{i+k+1}-t_{i+2}}\right]\right)$$

which is equal to

$$\frac{B_{i+k-1}}{t_{i+k}-t_i} - \frac{B_{i+1,k-1}}{t_{i+k+1}-t_{i+1}}$$
$$+\frac{k-1}{t_{i+k}-t_i}\left[\frac{x-t_i}{t_{i+k-1}-t_i}B_{i,k-2} + \frac{t_{i+k}-x}{t_{i+k}-t_{i+1}}B_{i+1,k-2}\right]$$
$$-\frac{k-1}{t_{i+k+1}-t_{i+1}}\left[\frac{x-t_{i+1}}{t_{i+k}-t_{i+1}}B_{i+1,k-2} + \frac{t_{i+k+1}-x}{t_{i+k+1}-t_{i+2}}B_{i+2,k-2}\right]$$

which implies the formula, using the Definition 1.3.2 of $B_{i,k-1}$ and $B_{i+1,k-1}$.

■

1.4 THE B-SPLINE BASIS OF $\mathcal{P}_{k,\tau,r}$

Let us now recall the notation of Section 1.1; k is a fixed positive integer, τ_i $(1 \leq i \leq \ell-1)$ is a sequence of points in $[a,b]$ such that $a < \tau_1 < \ldots < \tau_{\ell-1} < b$, and the elements of $\mathcal{P}_{k,\tau,r}$ are C^{r_i-1} at τ_i $(1 \leq i \leq \ell-1, \; 0 \leq r_i \leq k)$.

To be able to define B-splines as in the previous section, we have to define a sequence of knots t_i in such a way that the associated B-splines constitute a *basis* of the space $\mathcal{P}_{k,\tau,r}$.

1.4.1 Description of the knots (t_i)

Recall that a sequence τ_i $(1 \leq i \leq \ell-1, \tau_0 = a, \tau_\ell = b)$ of points in $[a,b]$ and a sequence $r = (r_i)$ $(1 \leq i \leq \ell-1)$ of integers with $0 \leq r_i \leq k$ are given.

Set $n = \ell(k+1) - \sum r_i$; n is the dimension of the space $\mathcal{P}_{k,\tau,r}$. Take a sequence of points (t_i) $(0 \leq i \leq m = n+k)$ with the following conditions:

(1) $t_i \leq t_{i+1}$

(2) $t_j \leq a$ $\;(0 \leq j \leq k)$

(3) $t_j \geq b$ $\;(n \leq j \leq n+k)$

(4) $(k+1-r_i)\, t_j$ are equal to τ_i, for all i $\;(1 \leq i \leq \ell-1)$.

Such t_i's are called *knots*, and they are uniquely determined by the sequences τ and r, except $(t_0, \ldots, t_k)$ and $(t_n, \ldots, t_{n+k})$ which only have to satisfy 2) and 3).

1.4.2 Definition   Let t_i $\;(0 \leq i \leq m)$ *be a sequence of points in* $\mathbf{R}$ *such that* $t_i \leq t_{i+1}$, k *an integer greater than or equal to 0, and* $[a,b]$ *an interval such that* $t_k \leq a$, $t_{m-k} \geq b$. *One denotes by* $\mathcal{P}_{k,t}([a,b])$ *(or simply* $\mathcal{P}_{k,t}$*) the vector space of piecewise polynomial functions on* $[a,b]$, *of degree* $\leq k$, *with* C^{k-p_j} *continuity at* t_j, *if* t_j *is a knot of multiplicity* p_j.

By assumption, a C^{k-p_j} continuity with $k - p_j < 0$ gives no condition at t_j.

With the relation between the (τ_i, r_i)'s and the t_j's defined above, one has $\mathcal{P}_{k,\tau,r} = \mathcal{P}_{k,t}$.

If all the knots are of multiplicity $\leq k+1$, we set $n = m - k$, and we have $\dim \mathcal{P}_{k,t} = n$ (1.4.1).

1.4.3 Theorem   *With the notation of 1.4.2 and 1.3.1, assume that all the knots are of multiplicity* $\leq k+1$; *then the* $B_{i,k,t}$ $(0 \leq i \leq n-1)$ *constitute a basis for the space* $\mathcal{P}_{k,t}$.

Proof   Recall first that by the hypothesis on the sequence (t_i), there is no knot of multiplicity $\geq k + 2$, so that none of the $B_{i,k,t}$'s is identically zero.

One has $n = \dim \mathcal{P}_{k,t}$; it is then enough to prove that, denoting by S the space generated by the $B_{i,k,t}$'s (in the space of the functions on $[a,b]$), we have $\mathcal{P}_{k,t} \subset S$. Recall (1.1.4) that a basis of $\mathcal{P}_{k,t}$ is given by

$$\begin{cases} (x-a)^{\nu} & (0 \leq \nu \leq k) \\ (x-\tau_i)_+^{\nu} & (1 \leq i \leq \ell - 1), \ (r_i \leq \nu \leq k); \end{cases}$$

each t_j is equal to some τ_i (1.4.1) and there are $(k + 1 - r_i)$ t_j equal to τ_i $(0 \leq r_i \leq k)$.

1.4.4 Lemma   *Without any hypothesis on the multiplicity of the knots, one has for* $t \in [a,b]$

a)
$$(x-t)^k = \sum_{i=0}^{n-1} \psi_{i,k}(t) B_{i,k}(x)$$

b)
$$(x-t_j)_+^k = \sum_{i \geq j} \psi_{i,k}(t_j) B_{i,k}(x) \quad for \quad 0 \leq j \leq n-1,$$

with $\psi_{i,k}(t) = (t_{i+1} - t) \ldots (t_{i+k} - t)$;
    $\psi_{i,k}$ is defined for $k \geq 1$, and is of degree k in t; one sets $\psi_{i,0}(t) = 1$ for all i.

Proof   b) is a consequence of a) because for $x \leq t_j$ and $i \geq j$, $B_{i,k}(x) = 0$, and then the two terms of b) are zero. For $x \geq t_j$ and $i < j$, either $\psi_{i,k}(t_j) = 0$ (for $i \geq j - k$), or $B_{i,k}(x) = 0$ (for $i < j - k - 1$), and so $\sum_{i=0}^{j-1} \psi_{i,k}(t_j) B_{i,k}(x) = 0$; we have now only to apply a).

Proof of a). We will argue by induction on k, the case $k = 0$ being clear. The induction hypothesis says that

$$\sum_{i=1}^{n-1} \psi_{i,k-1}(t) B_{i,k-1}(x) = (x-t)^{k-1}$$

for $x \in [a, b]$, as there are $k + 1$ knots smaller than or equal to a, one has $B_{0,k-1}(x) = 0$ for $x \in [a, b]$, which explains why the summation starts at $i = 1$.

We have:

$$\sum_{i=0}^{n-1} \psi_{i,k}(t) B_{i,k}(x)$$

$$= \sum_{i=0}^{n-1} \psi_{i,k}(t) \Big(\omega_{i,k}(x) B_{i,k-1}(x) + (1 - \omega_{i+1,k}(x)) B_{i+1,k-1}(x) \Big),$$

(by the definition of B-splines)

$$= \psi_{0,k}(t) \omega_{0,k}(x) B_{0,k-1}(x)$$

$$+ \sum_{i=1}^{n-1} B_{i,k-1}(x) \Big(\psi_{i,k}(t) \omega_{i,k}(x) + \psi_{i-1,k}(t)(1 - \omega_{i,k}(x)) \Big).$$

But if $t_i = t_{i+k}$, $B_{i,k-1}$ is $\equiv 0$, and if $t_i < t_{i+k}$, we have

$$\psi_{i,k}(t) \omega_{i,k}(x) + \psi_{i-1,k}(t)(1 - \omega_{i,k}(x))$$

$$= \psi_{i,k-1}(t)[(t_{i+k} - t)\omega_{i,k}(x) + (t_i - t)(1 - \omega_{i,k}(x))]$$

$$= \psi_{i,k-1}(t) \Big(\omega_{i,k}(x)[t_{i+k} - t_i] + t_i - t \Big)$$

$$= \psi_{i,k-1}(t)(x - t)$$

(Recall that $\psi_{i,k}(t) = (t_{i+1} - t) \ldots (t_{i+k} - t)$, and that $B_{0,k-1}(x) = 0$.)

We have therefore

$$\sum_{i=0}^{n-1} \psi_{i,k}(t) B_{i,k}(x) = (x - t) \sum_{i=1}^{n-1} \psi_{i,k-1} B_{i,k-1}(x) = (x - t)^k$$

by the induction hypothesis.

∎

Proof of Theorem 1.4.3

We have $(x - t)^k \in S$ by 1.4.4., differentiating formula a) with respect to t, we see that $(x - t)^{k-s} \in S$ $(0 \le s \le k)$:

$$(1) \qquad \frac{1}{(k - s)!}(x - t)^{k-s} = \frac{1}{k!} \sum_{i=0}^{n-1} (-D)^s \psi_{i,k}(t) B_{i,k}(x) \quad \forall t \in [a, b]$$

and in particular $(x - a)^{k-s} \in S$. Setting $t = t_j$ in (2), gives:

$$(2) \qquad \frac{1}{(k-s)!}(x - t_j)^{k-s} = \frac{1}{k!}\sum_{i=0}^{n-1}(-D)^s\psi_{i,k}(t_j)B_{i,k}(x).$$

Assume that there are $k + 1 - r$ knots t_i equal to t_j, for instance $t_j, \ldots, t_{j+k-r}$. Then we have

$$(-D)^{k-s}\psi_{i,k}(t_j) = 0 \text{ for } j - k \le i \le j - 1 \text{ and } 0 \le s \le k - r,$$

because $\psi_{i,k}$ has a zero of order $k+1-r$ at t_j, and $B_{i,k}(x) = 0$ for $i \le j-k-1$ and $x \ge t_j$, as the support of $B_{i,k}$ is $[t_i, t_{i+k+1}]$. We have therefore

$$\sum_{i=o}^{j-1}(-D)^s\psi_{i,k}(t_j)B_{i,k}(x) = 0 \quad \text{for} \quad 0 \le s \le k - r \text{ and } x \ge t_j.$$

One deduces then from (2)

$$\frac{1}{(k-s)!}(x - t_j)_+^{k-s} = \frac{1}{k!}\sum_{i\ge j}(-D)^s\psi_{i,k}(t_j)B_{i,k}(x)$$

for $0 \le s \le k - r$ and for all x, since for $x \le t_j$ the two terms are clearly equal to zero, which implies $(x - t_j)_+^\nu \in S$ $(r_j \le \nu \le k)$ as there are $k + 1 - r_j$ knots t_i equal to t_j, which completes the proof of 1.4.3.

∎

1.4.5 Remark

If some of the knots have multiplicity $\ge k + 2$, the above proof shows that the corresponding $B_{i,k,t}$'s span $\mathcal{P}_{k,t}$. They cannot be a basis, however, because if $t_i = \cdots = t_{i+k+1}$ (which corresponds to a knot of multiplicity $\ge k + 2$), $B_{i,k,t} \equiv 0$.

For the sake of completeness, let us now consider the integration of B-splines.

1.4.6 Proposition

$$\int_{-\infty}^{+\infty} B_{i,k}(x)\, dx = \frac{1}{k+1}(t_{i+k+1} - t_i).$$

Proof One may assume that $B_{i,k}$ has support in $[a, b]$, i.e., that $t_{i+k+1} \leq b$ (if not, one may replace b by $b' \geq t_{i+k+1}$, even if it means adding knots to the right of b', which leaves $B_{i,k}(x)$ unchanged). One may also add a knot $t_{-1} \leq t_0$, and a knot $t_{n+k+1} \geq t_{n+k}$, which does not modify $B_{i,k}(x)$, but allows one to consider the B-splines $B_{j,k+1}(x)$ $(x \in [a, b])$ for the knot sequence $t_{-1}, t_0, t_1, \ldots, t_{n+k+1}$.

Consider the indefinite integral $\int_{-\infty}^{x} B_{i,k}(t)\, dt$: it is an element of $\mathcal{P}_{k+1,t}$ which may be expressed as

$$\int_{-\infty}^{x} B_{i,k}(t)\, dt = \sum_{j=-1}^{n-1} c_j B_{j,k+1}(x)$$

for $x \in [a, b]$. Differentiating and applying 1.4.3 and 1.3.9, one gets

$$\begin{cases} c_0 = \cdots = c_{i-1} = 0 \\ c_i = \cdots = c_{n-1} = \dfrac{(t_{i+k+1} - t_i)}{k+1} \end{cases}$$

and therefore

$$\int_{-\infty}^{x} B_{i,k}(x)\, dx = \frac{(t_{i+k+1} - t_i)}{k+1} \left(\sum_{j \geq i} B_{j,k+1}(x) \right)$$

for $x \in [a, b]$, and $\int_{-\infty}^{x} B_{i,k}(x)\, dx = \frac{(t_{i+k+1} - t_i)}{k+1}$ for $t_{i+k+1} \leq x \leq b$ since then, by 1.3.4,

$$\sum_{j \geq i} B_{j,k+1}(x) = \sum_{j \geq -1} B_{j,k+1}(x) = 1.$$

■

1.5 BASIC ALGORITHMS FOR B-SPLINES

With the notation of the previous section, let

$$S(x) = \sum_{i=0}^{n-1} a_i B_{i,k}(x)$$

be an element of $\mathcal{P}_{k,t}$; we will call $S(x)$ a *spline function*, generalizing the cubic case of class $\mathcal{C}^2$ studied in Section 1.2.

We will use the notation of 1.3.1 and 1.4.2: the B-splines are computed with a knot sequence $t_0, \ldots, t_m$ and defined over all $\mathbf{R}$, and the algorithms described are independent of the chosen interval $[a, b]$ (so long as $t_k \leq a$ and $t_{m-k} \geq b$); in the algorithms described below, we have then set $n = m - k$.

One generally assumes that there is no knot of multiplicity $\geq k + 2$, which implies that the $B_{i,k,t}$'s ($0 \leq i \leq n - 1$) form a basis of $\mathcal{P}_{k,t}$ (by 1.4.3). However, this hypothesis is not indispensable, and is not in general satisfied for the spaces $\mathcal{P}_{k',t}$, $k' < k$ (which are considered in the inductions if there are multiple knots; for instance, one often has $t_0 = \cdots = t_k = a$ and $t_n = \cdots = t_{n+k} = b$).

We will describe three algorithms for the function $S(x)$: the "De Boor-Cox" or "De Casteljau" algorithm, allowing the evaluation of $S(x)$ at a point $x \in [a, b]$; the algorithm giving the coefficients (in relation to the $B_{i,k-1}$'s) of the derivative $S'(x)$; and finally the "knot insertion" algorithm, which computes the coefficients of $S(x)$ expanded in the basis $B_{i,k,t'}$ ($0 \leq i \leq n$), the knot sequence t' being obtained from the sequence t by adding a new knot $\hat{t}$.

a) Evaluation algorithm

1.5.1 Proposition   Let $x \geq t_k$, $S(x) = \sum_{i=0}^{n-1} a_i B_{i,k}(x)$. We have then

$$S(x) = \sum a_i^0 B_{i,k}(x) = \sum a_i^1(x) B_{i,k-1}(x) = \cdots = \sum a_i^k(x) B_{i,0}(x),$$

with

$$\begin{cases} a_i^0 = a_i \\ a_i^{r+1}(x) = \omega_{i,k-r}(x) a_i^r(x) + (1 - \omega_{i,k-r}(x)) a_{i-1}^r(x) \end{cases} ;$$

recall that

$$\omega_{i,k}(x) = \begin{cases} \dfrac{x - t_i}{t_{i+k} - t_i} & if \ t_i < t_{i+k} \\ 0 & (otherwise). \end{cases}$$

Proof   It is a direct application of the inductive definition of the B-splines; see 1.3.2.

∎

1.5.2 Remarks   1) The algorithm is also valid for $x < t_k$, with the convention that the B-splines of index ≤ 0 are indentically zero.

2) If one has to evaluate $S(x)$ for $x \in [t_j, t_{j+1}[$, one then sets $S(x) = a_j^k(x)$, as $B_{j,0}(x)$ is the characteristic function of the interval $[t_j, t_{j+1}[$. For the computation of $a_j^k(x)$, it is enough to evaluate a_i^r for $j - k + r \le i \le j$, the other $B_{i,k-r}(x)$ being identically zero, as $x \in [t_j, t_{j+1}[$ (see 1.3.4). We have therefore

$$S(x) = \sum_{i=j-k}^{j} a_i^0(x) B_{i,k}(x) = \sum_{i=j-k+1}^{j} a_i^1 B_{i,k-1}(x) = \cdots = a_j^k(x).$$

The algorithm is then "triangular", each element being a convex linear combination of the two above it:

$$
\begin{array}{cccccccc}
a_{j-k} & & a_{j-k+1} & \cdots & \cdots & \cdots & a_{j-1} & a_j \\
& a_{j-k+1}^1 & & \cdots & \cdots & \cdots & \cdots & a_j^1 \\
& & \ddots & & & & & \\
& & & & a_{j-1}^{k-1} & & a_j^{k-1} & \\
& & & & & a_j^k & &
\end{array}
$$

3) This algorithm is quite expensive: it requires the computation of $\frac{k(k+1)}{2}$ linear convex combinations, and for each one, it uses two multiplications, one division, and one substraction (for the computation of $1 - \omega_{i,k-r}(x)$).
Nevertheless, it has several advantages

 a) It is numerically stable.

 b) The computation of the $\omega_{i,k-r}(x)$'s is often very simple (for instance if x and the knots t_i are integers).

 c) It is well adapted to the tracing of spline curves (see the next chapter).

4) If one has to evaluate the function $S(x)$ at several points of $[t_i, t_{i+1}[$, one generally proceeds in another way.

 α) One computes once and for all the polynomial expression of S between t_i and t_{i+1}:

$$S(x) = \sum_{j=0}^{k} \frac{\alpha_j}{j!}(x - t_i)^j$$

with $\alpha_j = D^j(S)(t_i)$ (the right derivative evaluated with the differentiation algorithm, described below).

β) One evaluates $S(x)$ by the "Hörner rule" which needs only $2k$ operations.

Let us recall the Hörner rule. Let P be a polynomial of degree d: $P = a_d X^d + \cdots + a_1 X + a_0$; the program to evaluate P at x is

$$\begin{cases} P \leftarrow a_d \\ \text{For } i \ (d-1, 0, -1) \ \text{Do}: \ P \leftarrow (a_i + xP) \\ \text{End} \end{cases}$$

This amounts to writing $P(x)$ in the following way

$$P = x(\ldots(x(x a_d + a_{d-1}) + a_{d-2}) + \cdots + a_1) + a_0.$$

b) Differentiation algorithm

1.5.3 Proposition Let $S(x) = \sum_{i=0}^{n-1} a_i B_{i,k}(x)$. Then the right derivative $D(S(x))$ is given by the following formula

$$D(S(x)) = \sum_{i=1}^{n-1} b_i B_{i,k-1}(x)$$

with

$$b_i = \begin{cases} k\dfrac{(a_i - a_{i-1})}{(t_{i+k} - t_i)} & \text{if} \quad t_i < t_{i+k} \\ 0 & \text{otherwise.} \end{cases}$$

Proof It is enough to replace in $D(S(x))$, $D(B_{i,k}(x))$ by its expression (Proposition 1.3.9).

∎

Remark With the notation of Section 4, if $S(x) \in \mathcal{P}_{k,t}$, then $D(S(x)) \in \mathcal{P}_{k-1,t}$, and one always has $B_{0,k-1}|[a,b] \equiv 0$, which explains why it is that in $D(S(x))$, the sum goes from 1 to $n-1$.

c) Knot insertion algorithm

Let $S(x) = \sum_{i=0}^{n-1} a_i B_{i,k}(x)$ be a spline function defined with a knot sequence $(t_i)_{0 \le i \le m=n+k}$ and let us add a new knot $\hat{t}$ to the sequence t_i such that $\hat{t} \le t_{n-1}$ ($\hat{t}$ may be equal to one of the t_i's). To the new sequence so obtained (denoted t') there corresponds a set of B-splines $(\hat{B}_{i,k})$ $0 \le i \le n$, and one has $\mathcal{P}_{k,t} \subset \mathcal{P}_{k,t'}$.

1.5.4 Proposition   *One has*

$$\sum_{i=0}^{n-1} a_i B_{i,k}(x) = \sum_{i=0}^{n} \hat{a}_i \hat{B}_{i,k}(x)$$

with

$$\hat{a}_i = \begin{cases} a_i & \text{if } t_{i+k} \leq \hat{t} \\ \omega_{i,k}(\hat{t})a_i + (1 - \omega_{i,k}(\hat{t}))a_{i-1} & \text{if } t_i < \hat{t} < t_{i+k} \\ a_{i-1} & \text{if } \hat{t} \leq t_i. \end{cases}$$

Proof   One may argue by induction on k, applying the first step of algorithm 1.5.1 to the two terms, and the induction hypothesis. Let $x \in \mathbf{R}$. One has, by 1.5.1,

$$\sum_{i=0}^{n-1} a_i B_{i,k}(x) = \sum_{i=0}^{n-1} a_i^1 B_{i,k-1}(x)$$

$$= \sum_{i=0}^{n} (a_i^1) \hat{} \hat{B}_{i,k-1}(x)$$

(by the induction hypothesis), and, again by 1.5.1,

$$\sum_{i=0}^{n} \hat{a}_i \hat{B}_{i,k}(x) = \sum_{i=0}^{n} (\hat{a}_i)^1 \hat{B}_{i,k-1}(x).$$

It is then enough to see that $(a_i^1) \hat{} = (\hat{a}_i)^1$. One has, by 1.5.1,

$$a_i^1 = \left(\frac{x - t_i}{t_{i+k} - t_i}\right) a_i + \left(\frac{t_{i+k} - x}{t_{i+k} - t_i}\right) a_{i-1}$$

and

$$\sum_{i=0}^{n-1} a_i^1 B_{i,k-1}(x) = \sum_{i=0}^{n} (a_i^1) \hat{} B_{i,k-1}(x), \text{ with}$$

$$(a_i^1) \hat{} = \begin{cases} a_i^1 & \text{if } t_{i+k-1} \leq \hat{t} \\ \dfrac{\hat{t} - t_i}{t_{i+k-1} - t_i}a_i^1 + \dfrac{t_{i+k-1} - \hat{t}}{t_{i+k-1} - t_i}a_{i-1}^1 & \text{if } t_i < \hat{t} < t_{i+k-1} \\ a_{i-1}^1 & \text{if } \hat{t} \leq t_i \end{cases}$$

by the induction hypothesis.

If i is chosen such that $t_i < \hat{t} < t_{i+k-1}$, one has

$$(a_i^1)\hat{} = \frac{\hat{t} - t_i}{(t_{i+k-1} - t_i)} \frac{x - t_i}{(t_{i+k} - t_i)} a_i + \frac{\hat{t} - t_i}{(t_{i+k-1} - t_i)} \frac{t_{i+k} - x}{(t_{i+k} - t_i)} a_{i-1}$$
$$+ \frac{t_{i+k-1} - \hat{t}}{(t_{i+k-1} - t_i)} \frac{x - t_{i-1}}{(t_{i+k-1} - t_{i-1})} a_{i-1}$$
$$+ \frac{t_{i+k-1} - \hat{t}}{(t_{i+k-1} - t_i)} \frac{t_{i+k-1} - x}{(t_{i+k-1} - t_{i-1})} a_{i-2}.$$

Let us now compute $\sum_{i=0}^n \hat{a}_i \hat{B}_{i,k}(x)$ with $\hat{a}_i$ expressed as a function of the a_i's as in 1.5.4:

$$\hat{a}_i = \left(\frac{\hat{t} - t_i}{t_{i+k} - t_i}\right) a_i + \left(\frac{t_{i+k} - \hat{t}}{t_{i+k} - t_i}\right) a_{i-1}$$

(assuming always that $t_i < \hat{t} < t_{i+k-1}$). One has

$$(\hat{a}_i)^1 = \left(\frac{x - t_i}{t_{i+k-1} - t_i}\right) \hat{a}_i + \left(\frac{t_{i+k-1} - x}{t_{i+k-1} - t_i}\right) \hat{a}_{i-1}$$

by 1.5.1 applied to the sequence t', because, since there is a new knot between t_i and t_{i+k-1} in the sequence t', one has $\hat{t}_{i+k} = t_{i+k-1}$. This implies

$$(\hat{a}_i)^1 =$$
$$\left(\frac{x - t_i}{t_{i+k-1} - t_i}\right)\left(\frac{\hat{t} - t_i}{t_{i+k} - t_i}\right) a_i + \left(\frac{x - t_i}{t_{i+k-1} - t_i}\right)\left(\frac{t_{i+k} - \hat{t}}{t_{i+k} - t_i}\right) a_{i-1}+$$

$$\left(\frac{t_{i+k-1} - x}{t_{i+k-1} - t_i}\right)\left(\frac{\hat{t} - t_{i-1}}{t_{i+k-1} - t_{i-1}}\right) a_{i-1} + \left(\frac{t_{i+k-1} - x}{t_{i+k-1} - t_i}\right)\left(\frac{t_{i+k-1} - \hat{t}}{t_{i+k-1} - t_{i-1}}\right) a_{i-2},$$

and it is easy to check that $(a_i^1)\hat{} = (\hat{a}_i)^1$. The verification of 1.5.4 for the other values of i (i.e., such that one does not have $t_i < \hat{t} < t_{i+k-1}$) is left to the reader. ∎

As an application, let us prove the "variation diminishing property" for spline functions.

1.5.5 Definition   *1) Let $\underline{a} = (a_0, \ldots, a_{n-1})$ be a sequence of real numbers; the variation of $\underline{a}$, noted $V(\underline{a})$, is by definition the number of sign changes* of the sequence $\underline{a}$, i.e., the number of indices i such that $a_i.a_{i+r} < 0$, with $a_{i+1} = \cdots = a_{i+r-1} = 0$.

For instance, for the sequence $\underline{a} = (-1, 0, 0, 2, 0, 4, 0, 0, -5, 6)$, one has $V(\underline{a}) = 3$.

2) Let $f : [a, b] \longrightarrow \mathbf{R}$ be a function; set $V(f)$ for the variation of f, or the number of change of signs of f, defined as $\operatorname{Sup}V(f(x_1), \ldots, f(x_r))$ for all sequences of points $x = (x_1, \cdots, x_r)$ $(x_1 < \ldots < x_r)$ in $[a, b]$.

One has then, with the previous notation

1.5.6 Proposition ("Variation diminishing property for spline functions")

Let $S(x) = \sum_{i=0}^{n-1} a_i B_{i,k}(x)$ be a spline function defined over an interval $[a, b]$ such that $t_k \leq a$ and $t_n \geq b$. Then $V(S) \leq V(\underline{a})$.

Let us first prove a lemma.

With the notation of Proposition 1.5.4 (knot insertion algorithm), let us add a new knot $\hat{t}$ to the sequence t_i, and let $\underline{\hat{a}} = (\hat{a}_i)_{1 \leq i \leq n+1}$ be the sequence of coefficients of the function $S(x)$ in the basis $\hat{B}_{i,k}$. We have then

1.5.7 Lemma

$$V(\underline{\hat{a}}) \leq V(\underline{a}).$$

Proof Each $\hat{a}_i$ is a linear convex combination of a_{i-1} and a_i (see 1.5.4); if then we add $\hat{a}_j$ between a_{j-1} and a_j, the variation of the sequence remains constant. For in passing from the sequence $(\underline{a})$ to the sequence $(\underline{\hat{a}})$, we may begin by adding, for each index i, $\hat{a}_i$ between a_{i-1} and a_i, which does not change the variation, and remove the a_j's after, which may only lower the variation.

∎

Proof of 1.5.6 Let $x_1, \ldots, x_r$ $(x_i < x_{i+1})$ be r points in $[a, b]$; we have to show that $V(S(x_1), \ldots, S(x_r)) \leq V(\underline{a})$. Let us add knots equal to x_1 to the sequence t_i, until we obtain a knot of multiplicity k at x_1 (if x_1 is not equal to a knot t_i, we add k knots equal to x_1; we may in fact always assume that this is the case).

We obtain in this way a new sequence $(\tilde{a}_i^1)_{0 \leq i \leq n+k-1}$ of coefficients for the function $S(x)$, corresponding to B-splines $\tilde{B}_i^1$, and such that $V(\tilde{a}_i^1) \leq V(a_i)$, as follows from Lemma 1.5.7.

There is then a knot of multiplicity k at x_1: there exists therefore an index j such that $\tilde{B}_j^1(x_1) = 1$ (see 1.3.4, e)), and so $S(x_1)$ is an element of the sequence $(\tilde{a}_i^1)$; proceeding in the same way for all the x_i's, one gets a sequence $(\tilde{a}_i^r)$ such that $V(\tilde{a}_i^r) \leq V(a_i)$, and containing $S(x_i)$ $(1 \leq i \leq r)$; removing then the superfluous $\tilde{a}_i^r$, which may only diminish the variation, we obtain $V\big(S(x_1), \ldots, S(x_r)\big) \leq V(\underline{a})$.

∎

1.5.8 Remarks 1) Proposition 1.5.6 implies that if we assume that the roots of $S(x)$ are simple, then the number of zeros of $S(x)$ in $[a, b]$ is bounded by $V(a_i)$. Therefore 1.5.6 may be considered as a generalization to spline functions, of *Descartes's lemma* (or *Descartes's rule of signs*) which applies to real polynomials.

Let us recall this lemma

1.5.9 Lemma Let $P = \sum_{i=0}^r a_i X^i \in \mathbf{R}[X]$. If $Z_+(P)$ designates the number of roots > 0 of P (counted with multiplicity), then $Z_+(P) \leq V(\underline{a})$.

The proof is easy by induction on r, using Rolle's lemma (see [B-R]). In fact, there exists a proof of 1.5.6 using a generalization of Rolle's lemma to spline functions: see [Sc].

2) Using Proposition 1.5.3, which computes the derivative of $S(x) = \sum_{i=0}^{n-1} a_i B_{i,k}$, one may easily prove the following result:

1.5.10 Corollary ("Monotony diminishing") If $S(x)$ is of class $\mathcal{C}^1$, then $V(S'(x)) \leq V(a_i - a_{i-1})$ $(1 \leq i \leq n-1)$.

∎

1.6 APPROXIMATION BY A SPLINE CURVE

Let $y = g(x)$ be a curve γ of class $\mathcal{C}^2$ defined on $[a, b]$, $(t_i)_{0 \leq i \leq n+k}$ a sequence of knots as in the previous section (i.e., such that $t_k \leq a$ and $t_{n+1} \geq b$, allowing us to define B-splines $B_{i,k}$ of degree k $(k \geq 1$, $0 \leq i \leq n-1)$. Set $t_i^* = \frac{t_{i+1} + \cdots + t_{i+k}}{k}$ and $Sg(x) = \sum_{i=0}^{n-1} g(t_i^*) B_{i,k}(x)$. Then $y = Sg(x)$ is a spline curve defined on $[a, b]$, which will be the curve approximating γ, and will be denoted $S\gamma$.

Let us first give a property of the operator S (called Schoenberg's operator) which motivates the choice of the sequence t^*.

1.6.1 Lemma   *The operator S "reproduces lines", i.e., one has $S\ell = \ell$ for any polynomial $\ell(X)$ of degree ≤ 1.*

Proof   We have to show that $S1 = 1$ and $SX = X$; the former comes from 1.3.4, d); for the latter, proving $(SX)' = 1$ is enough, which is easy using the differentiation algorithm 1.5.3 above.

∎

1.6.2 Proposition   *Set $|t| = \sup_{k \leq i \leq n-1}(t_{i+1} - t_i)$. Then for $k \geq 1$, there exists a constant $C(k)$ $(C(k) = \frac{k^2}{2})$ such that*

$$\|g - Sg\| \leq C(k)|t^2| \, \|g''\|,$$

the norm being the uniform norm on $[a,b]$, namely, $\|f\| = \sup_{x \in [a,b]} |f(x)|$.

Proof   Let $x_0 \in]a, b[$; we have to estimate $|g(x_0) - Sg(x_0)|$.

Assume $x_o \in [t_i, t_{i+1}[$, which implies $t^*_{i-k} \leq x_0 < t^*_i$, and $B_{j,k}(x_0) = 0$ for $j \notin [i-k, i]$.

Let $P(x) = g(x_0) + g'(x_0)(x - x_0)$ be the equation of the tangent to γ at x_0. We have $SP = P$ by 1.6.1, so

$$S(g - P) = Sg - SP = Sg - P = (Sg - g) + g - P$$

and, for $x = x_0$,

$$g(x_0) - Sg(x_0) = -S(g - P)(x_0) = -\sum_{i-k}^{i}(g - P)(t^*_j)B_{j,k}(x_0).$$

This implies

$$|g(x_0) - Sg(x_0)| \leq Sup|(g - P)(x)| \quad (t^*_{i-k} \leq x < t^*_i);$$

but $(g - P)(x) = \frac{(x-x_0)^2}{2}g''(\xi)$ with $\xi \in]x_0, x[$ by Taylor's formula, and $|x - x_0| \leq k|t|$ for $x \in [t^*_{i-k}, t^*_i[$, and the formula

$$|g(x_0) - Sg(x_0)| \leq \frac{k^2}{2}|t^2| \, \|g''\|.$$

∎

The curve $S\gamma : y = Sg(x)$ has an additional property, linked to the "variation diminishing property" of the spline curves (1.5.6): it is *regularizing*, that is, for any line D, $\mathrm{card}(D \cap S\gamma) \leq \mathrm{card}(D \cap \gamma)$ (see 1.6.4 below).

1.6.3 Remark   If g is of class C^r, $r > 2$ one may, by replacing the t_i^*'s by appropriate expressions in the t_i's, approximate γ to the order $|t|^r$ by spline curves (see [Sc]); but then one no longer has the property that Sg is regularizing.

1.6.4 Proposition   Let γ be a curve of class C^1 defined on $[a, b]$ by the equation $y = g(x)$. Then for any line D transversal to $S\gamma$, one has

$$\mathrm{card}(S\gamma \cap D) \leq \mathrm{card}(\gamma \cap D).$$

Proof
The transformation S reproduces lines, i.e., $S\ell = \ell$ for any affine function $\ell(x)$ (see 1.6.1). Let $y = \ell(x)$ be the equation of D; since D intersects $S\gamma$ transversally, one has $\mathrm{card}(S\gamma \cap D) \leq V(Sg - \ell)$, V being the variation in the interval $[a, b]$ (see Figure 1.6.1). However,

$$
\begin{aligned}
V(Sg - \ell) &= V\big(S(g - \ell)\big) \quad \text{(as } S\ell = \ell\text{)}\\
&= V\left(\sum_{i=0}^{n-1}(g - \ell)(t_i^*)B_i(x)\right)\\
&\leq V\big((g - \ell)(t_i^*)\big) \quad \text{(by 1.5.6)}\\
&\leq \mathrm{card}(\gamma \cap D).
\end{aligned}
$$

∎

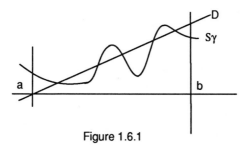

Figure 1.6.1

1.7 DIVIDED DIFFERENCES

B-splines are classically defined with "divided differences".

We will recall their definition and some properties, for the sake of completeness, and to help the reader to make connection with classical books on B-splines. Moreover, this definition will allow us to generalize B-splines to several variables.

The proofs will often be only sketched. One may look for instance at [DB] for more details.

1.7.1 Notation

a) If $t_0 < \cdots < t_r$ are points of **R** and $g_0, \ldots, g_r$ functions, one sets

$$D\begin{pmatrix} t_0, \ldots, t_r \\ g_0, \ldots, g_r \end{pmatrix}$$

the determinant of the matrix $(r+1) \times (r+1)$: $(g_j(t_i))$. (Rows and columns are numbered from 0 to r).

b) If $t_0 \leq \ldots \leq t_r$, and if $t_0 = t_1 = \cdots = t_{\ell-1} < t_\ell$, one denotes in the same way the determinant of the matrix $(D^{d_i} g_j(t_i))$, where $d_i = \sup\{j : t_i = t_{i-1} = \cdots = t_{i-j}\}$, i.e., of the matrix

$$\begin{pmatrix} g_0(t_0) & g_1(t_0) & \cdots & \cdots \\ g_0'(t_0) & \cdots & \cdots & \cdots \\ \vdots & & & \\ g_0^{(\ell-1)}(t_0) & \cdots & \cdots & \cdots \\ g_0(t_\ell) & \cdots & \cdots & \cdots \\ \vdots & & & \end{pmatrix}$$

(and in the same way for the other t_i's).

1.7.2 Definition   Let $t_0 \leq \cdots \leq t_r$ be points of **R** and f a function sufficiently differentiable. One defines the r-th divided difference of f, denoted $[t_0, \ldots, t_r]f$, in one of the two following equivalent ways:

a) $[t_0, \ldots, t_r]f$ is the coefficient of X^r in the unique polynomial P of degree $\leq r$ such that $P(t_i) = f(t_i)$ for $0 \leq i \leq r$ (and $P^{(j)}(t_i) = f^{(j)}(t_i)$, $0 \leq j \leq \ell_i - 1$, if there are ℓ_i points t_j equal to t_i).

b)

$$[t_0, \ldots t_r]f \; = \; \frac{D\begin{pmatrix} t_0, \ldots\ldots\ldots, t_r \\ 1, \ldots, X^{r-1}, f \end{pmatrix}}{D\begin{pmatrix} t_0, \ldots, t_r \\ 1, \ldots, X^r \end{pmatrix}}$$

(see 1.7.1 for notation).

Let us sketch the proof of the equivalence. It is enough to prove by induction on r, that definitions a) and b) satisfy the following properties

$$(1.7.1, \; \alpha) \qquad\qquad [t_i]f \;\; = \;\; f(t_i)$$

$$(1.7.1, \; \beta) \quad [t_0, \ldots, t_r]f = \begin{cases} \dfrac{f^{(r)}(t_0)}{r!} & \text{if } t_0 = \cdots = t_r \\[2ex] \dfrac{[t_1, \ldots t_r]f - [t_0, \ldots t_{r-1}]f}{t_r - t_0} & \text{if } \; t_0 < t_r. \end{cases}$$

For definition a), the reader may look at [D B], chapter 1, and for definition b), prove that

$$[t_0, \ldots, t_r]X^i \;\; = \;\; \begin{cases} 0 & \text{for } \; i < r \\ 1 & \text{for } \; i = r \\ t_0 + \cdots + t_r & \text{for } \; i = r+1, \end{cases}$$

and consider the linear form which in the case $t_0 < t_r$ is defined by

$$\lambda(f) = [t_0, \ldots, t_r]f - \frac{[t_1, \ldots, t_r]f - [t_0, \ldots, t_{r-1}]f}{(t_r - t_0)}.$$

One then has $\lambda(X^i) = 0$ for $0 \le i \le r$, and so if $P_r(f)$ is the polynomial of degree $\le r$ such that $f(t_i) = P(t_i)$ $(0 \le i \le r)$ with the above convention (i.e., considering derivatives in case of multiple t_i's), one has $\lambda(P_r(f)) = 0$, so $\lambda(f - P_r(f)) = \lambda f$ and also $\lambda(f - P_r(f)) = 0$, since the function $f - P_r(f)$ is zero at the t_i's.

1.7.3 Proposition   *Let k be an integer ≥ 1, $(t_i)_{0 \le i \le n+k}$, $t_i \le t_{i+1}$ a sequence of points as in 1.3.1. We then have*

$$B_{i,k,t}(x) \;\; = \;\; (t_{i+k+1} - t_i)[t_i, \ldots, t_{i+k+1}] \, (. - x)_+^k,$$

the notation $[t_i, \ldots, t_{i+k+1}] \, (. - x)_+^k$ meaning that one applies the divided difference operator $[t_i, \ldots, t_{i+k+1}]$ to the function $y \mapsto (y - x)_+^k$.

Proof   We have to prove that the function defined in 1.7.3 satisfies the induction relation 1.3.2. We use the following lemma.

1.7.4 Lemma   ("Leibniz's formula") *Let f, g, h be three functions defined on the interval $[t_0, t_r]$ such that $f = gh$; then*

$$[t_0, \ldots, t_r]f = \sum_{i=0}^{r}([t_0, \ldots, t_i]g)\,([t_i, \ldots, t_r]h).$$

For a proof, see [DB] p. 5.

To prove 1.7.3, we now argue by induction on k, the case $k = 0$ being easy because $(t_{i+1} - t_i)[t_i, t_{i+1}]f = f(t_{i+1}) - f(t_i)$, and we apply that to $h(y) = (y - x)_+^0$ which is 0 for $y \leq x$ and 1 for $y > x$; the fact $h(x) = 0$ is a convention, taken to make the spline functions right differentiable, see 1.3.4.

For the general case $(k \geq 1)$, we apply Leibniz's formula 1.7.4 to $f = (y - x)_+^k = (y - x)(y - x)_+^{k-1}$, and use the induction hypothesis.

∎

1.7.5 Hermite-Genocchi formula

Let $S_k = \{(\tau_0, \ldots, \tau_k),\ \tau_i \geq 0,\ \sum_{i=0}^{k} \tau_i = 1\}$,the k-dimensional simplex (embedded in $\mathbf{R}^{k+1}$). We then have, for any C^k function g

$$[t_0, \ldots, t_k]g = \int_{S_k} g^{(k)}(t_0\tau_0 + \cdots + t_k\tau_k)d\tau_1 \ldots d\tau_k$$

(the notation $\int_{S_k}$ means that we have set $\tau_0 = 1 - \tau_1 - \cdots - \tau_k$ and we integrate over the domain $\tau_i \geq 0$ $(1 \leq i \leq k)$ and $\sum_{i=1}^{k} \tau_i \leq 1$).

Proof

1) The formula is true if $t_0 = \cdots = t_k$, because

$$\int_{S_k} g^{(k)}(t_0)d\tau_1 \ldots d\tau_k = g^{(k)}(t_0) \int_{S_k} d\tau_1 \ldots d\tau_k = g^{(k)}(t_0)/k!$$

since the volume of S_k is $\frac{1}{k!}$, which is easily seen by induction on k.

2) Assume $t_0 < t_k$, and let us prove the formula by induction on k. If $k = 1$, we have:

$$\int_0^1 g'\big(t_0 + \tau_1(t_1 - t_0)\big)d\tau_1 = \frac{g(t_1) - g(t_0)}{t_1 - t_0} = [t_0, t_1]g.$$

To pass from $k - 1$ to k, one writes

$$\int_{S_k} g^{(k)}(t_0\tau_0 + \cdots + t_k\tau_k)d\tau_1 \ldots d\tau_k$$

$$= \int_{S_k} g^{(k)}\big(t_0 + (t_1 - t_0)\tau_1 + \cdots + (t_k - t_0)\tau_k\big)d\tau_1 \ldots d\tau_k$$

$$= \int_{S_{k-1}} \left[\int_0^{1-\sum_{i=1}^{k-1} \tau_i} g^{(k)}\big(t_0 + (t_1 - t_0)\tau_1 + \cdots + (t_k - t_0)\tau_k\big) d\tau_k \right] d\tau_1 \ldots d\tau_{k-1}$$

$$= \frac{1}{t_k - t_0} \Big([t_1, \ldots, t_k]g - [t_0, \ldots, t_{k-1}]g \Big)$$

by the induction hypothesis, and this last expression is equal to $[t_0, \ldots, t_k]g$ by property (1.7.1, β) above. ∎

1.7.6 Corollary *Let g be a C^1 function on $\mathbf{R}$ with compact support. One then has the formula*

$$\int_{-\infty}^{+\infty} g(x) B_{i,k,t}(x) dx$$

$$= k!\,(t_{i+k+1} - t_i) \int_{S_{k+1}} g(t_i \tau_0 + \cdots + t_{i+k+1}\tau_{k+1}) d\tau_1 \ldots d\tau_{k+1}$$

Proof Let G be a function sufficiently differentiable, such that $G^{(k+1)} = g$. We then have

$$\int_{S_{k+1}} g(t_i \tau_0 + \cdots + t_{i+k+1}\tau_{k+1}) d\tau_1 \ldots d\tau_{k+1} = [t_i, \ldots, t_{i+k+1}]G$$

by 1.7.5. It is enough to prove that

$$\int_{-\infty}^{+\infty} g(x) B_{i,k,t}(x) dx = k!\,(t_{i+k+1} - t_i)[t_i, \ldots, t_{i+k+1}]G.$$

This we will do by induction on k, the formula being clearly true for $k = 0$ (with the usual convention $0! = 1$). We have, denoting by G_1 a primitive of g,

$$\int_{-\infty}^{+\infty} g(x) B_{i,k,t}(x) dx = - \int_{-\infty}^{+\infty} G_1(x) B'_{i,k,t}(x) dx$$

(integration by parts)

$$= - \int_{-\infty}^{+\infty} G_1(x) k \left[\frac{B_{i,k-1}}{t_{i+k} - t_i} - \frac{B_{i+1,k-1}}{t_{i+k+1} - t_{i+1}} \right] dx \quad \text{by 1.3.9}$$

$$= k!\,(t_{i+k+1} - t_i)[t_i, \ldots, t_{i+k+1}]G$$

by the induction hypothesis and formula (1.7.1, β). ∎

The formula of Corollary 1.7.6 may be generalized to the case where x is a variable of $\mathbf{R}^s$, which defines $B_{i,k}(x)$ as a "distribution" on $\mathbf{R}^s$, generalizing the one-variable B-splines. These "B-splines" in several variables will be defined and studied in Section 4.7 of Chapter 4.

2

Spline curves and Bézier curves

2.1 BERNSTEIN POLYNOMIALS

We will introduce Bernstein polynomials as special cases of B-splines; Bézier curves will then appear as special spline curves, and will be treated specifically only as examples.

Let us fix k, and set $[a, b] = [0, 1]$; we will look at the B-splines $B_{i,k,t}$, the knot sequence $t = (t_i)$ being defined on the following way

$$\begin{cases} t_0 = \cdots = t_k = 0 \\ t_{k+1} = \cdots = t_{2k+1} = 1. \end{cases}$$

The $B_{i,k}(X)$'s $(0 \le i \le k)$ are then polynomials of degree k on $[0,1]$, verifying after 1.3.2

$$B_{i,k}(X) = X B_{i,k-1}(X) + (1 - X) B_{i+1,k-1}(X).$$

The $B_{i,k}$'s then make a basis of the space of polynomials (on $[0,1]$) of degree $\le k$ (1.4.2), called *the Bernstein basis*.

2.1.1 Notation

Traditionally, the Bernstein polynomials $B_{i,k}$ are denoted B_i^k, which makes a distinction with the general B-splines. We will use this notation when speaking of Bézier polynomials.

2.1.2 Proposition   *We have*

$$B_i^k(X) = \binom{k}{i}(1-X)^{k-i}X^i \quad (0 \le i \le k).$$

Proof   The induction relation of B-splines is

(2.1.2, 1) $$B_{i,k,t}(X) = X B_{i,k-1,t}(X) + (1-X)B_{i+1,k-1,t}(X)$$

where $B_{i,k}$ is the B-spline of degree k defined with the knot sequence

$$(t) \quad \begin{cases} t_0 = \cdots = t_k = 0 \\ t_{k+1} = \cdots = t_{2k+1} = 1. \end{cases}$$

On the other hand, the Bernstein polynomial B_j^{k-1} is by definition equal to $B_{j,k-1,t'}$, the spline $B_{j,k-1,t'}$ being now evaluated with the knot sequence

$$(t') \quad \begin{cases} t'_0 = \cdots = t'_{k-1} = 0 \\ t'_k = \cdots = t'_{2k-1} = 1. \end{cases}$$

We have then $B_{j,k-1,t} = B_{j-1,k-1,t'} = B_{j-1}^{k-1}$ $(1 \le j \le k)$, the spline $B_{0,k-1,t}$ being identically zero (as there is a knot of multiplicity $k+1$ at 0, for the sequence (t)); the relation (2.1.2, 1) becomes then (for Bernstein polynomials)

$$B_i^k(X) = X B_{i-1}^{k-1}(X) + (1-X)B_i^{k-1}(X),$$

and so $B_i^k(X) = \binom{k}{i}(1-X)^{k-i}X^i$, after the formula: $\binom{k}{i} = \binom{k-1}{i} + \binom{k-1}{i-1}$. ∎

Let us now prove a proposition about properties of the Bernstein basis, similar to proposition 1.3.2 for B-splines.

2.1.3 Proposition

a) The $B_i^k(X)$'s $(0 \le i \le k)$ constitute a basis of the space $\mathcal{P}_k$ (polynomials of degree $\le k$).
b) $B_i^k(X) \ge 0$ $(0 \le X \le 1)$.
c) $\sum_{i=0}^k B_i^k(X) = 1$.
d) If one sets $f(X) = \sum_{i=0}^n a_i B_i^k(X)$, one has $V(f) \le V(a_i)$.
e) $D B_i^k(X) = k\big(B_{i-1}^{k-1}(X) - B_i^{k-1}(X)\big)$.

f) $\forall X \in [0, 1]$, one has $B_i^k(X) = B_{k-i}^k(1 - X)$, $\quad (0 \le i \le k)$.

Proof

These conditions are similar to those for B-splines (see 1.3.4, 1.4.3, and 1.5.6), except for condition f). The direct proof of 2.1.3 is anyway very easy.

∎

2.1.4 Example

Let us represent the elements $B_i^3(X)(0 \le i \le 3)$ of the Bernstein basis in degree 3:

$$B_0^3(X) = (1 - X)^3, \quad B_1^3(X) = 3(1 - X)^2 X,$$
$$B_2^3(X) = 3(1 - X)X^2, \quad B_3^3(X) = X^3.$$

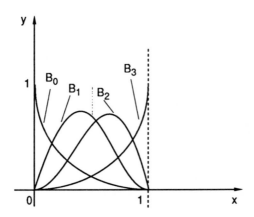

Figure 2.1.1

2.1.5 Remark In the case of any interval $[a, b]$, with

$$\begin{cases} t_0 = \cdots = t_k = a \\ t_{k+1} = \cdots = t_{2k+1} = b, \end{cases}$$

the corresponding $B_{i,k,t}(X)$'s $\quad (0 \le i \le k)$ constitute the *Bernstein basis* for the interval $[a, b]$; if $B_i^k(X)$ $\quad (0 \le i \le k)$ is the Bernstein basis on $[0, 1]$, we then have $B_{i,k,t}(X) = B_i^k\left(\frac{X-a}{b-a}\right)$ $\quad (a \le x \le b)$, as is easily seen.

2.2 B-SPLINES CURVES

We will denote by t the variable of a parametric curve. Apart from this change in the name of the variable, the notation remains the same as in Chapter 1, Section 1.3.

Let $P_0, \ldots, P_{n-1}$ be n points in $\mathbf{R}^s$.

2.2.1 Definition   a) *The B-spline curve (one also says "spline curve") associated to the polygon $P_0, \ldots, P_{n-1}$, is the parametric curve γ defined by*

$$S(t) = \sum_{i=0}^{n-1} P_i B_{i,k}(t) \qquad (a \leq t \leq b).$$

The polygon $P_0, \ldots, P_{n-1}$ is called the control polygon of the curve γ.

b) *In the special case of Bernstein polynomials $B_i^k(t)$, the curve*

$$B(t) = \sum_{i=0}^{k} P_i B_i^k(t)$$

is called the Bézier curve associated to the polygon $(P_0, \ldots, P_k)$; in this case, the number of P_i's is equal to $k+1$, if we look at Bernstein polynomials of degree k.

The properties of B-spline functions studied in the preceding section give the following

2.2.2 Proposition

a) γ *does not in general pass through the points P_i; one has however $S(a) = P_0$ and $S(b) = P_{n-1}$ when $t_0 = \cdots = t_k = a$, $t_n = \cdots = t_{n+k} = b$, and in this case, γ is tangent at P_0 and P_{n-1} to the sides of the control polygon.*

b) γ *is in the convex hull of the points $P_0, \ldots, P_{n-1}$. To be more precise, if $t_i \leq t < t_{i+1}$, $S(t)$ is in the convex hull of $(P_{i-k}, \ldots, P_i)$.*

c) *When the knots t_i $(k+1 \leq i \leq n-1)$ are simple, γ is a C^{k-1} curve, and is formed with n paramatrized polynomial arcs of degree $\leq k$.*

d) γ *is invariant by affine maps of $\mathbf{R}^s$: if h is such a map, the points $h(P_i)$ are the control points of the curve $h\big(S(t)\big)$.*

e) $S(t)$ *is regularizing, that is, if $\mathbf{P}$ denotes the control polygon of γ, and if H is a transversal hyperplane to γ, one has*

$$\operatorname{card}(H \cap \gamma) \leq \operatorname{card}(H \cap \mathbf{P}).$$

Proof

a) Easy, using 1.5.3 for the evaluation of the derivatives at a and b.

b) Comes from the relation $\sum_{i=0}^{n-1} B_{i,k}(t) = 1$ and from the fact that if $t_i \leq t < t_{i+1}$, one has $B_{j,k}(t) = 0$ for $j \leq i - k - 1$ and $j \geq i + 1$ (see 1.3.4).

c) Clear by definition.

d) Using the relation $\sum_{i=0}^{n-1} B_{i,k}(t) = 1$, we see that if h is an affine function: $\mathbf{R}^s \longrightarrow \mathbf{R}$, we have

$$h \circ S(t) = \sum_{i=0}^{n-1} h(P_i) B_{i,k}(t),$$

which implies d).

e) Let $h = 0$ be an equation of H. If $\sharp$ denotes the cardinal of a set, we have $\sharp(H \cap \gamma) = V(h \circ S(t))$ (as H meets γ transversally), V being the variation (1.5.5). Then $V(h \circ S(t)) \leq V(h(P_i))$ by d) and 1.5.6, and $V(h(P_i)) = \sharp(\gamma \cap \mathbf{P})$.

■

2.2.3 Remark   The B-spline curves have a *local* behavior, in the following sense:

a) If X is a point on the curve with parameter value t_0, the position of X depends only on at most $k + 1$ points P_i, since if $t_j \leq t_0 < t_{j+1}$, we have

$$S(t_0) = \sum_{i=0}^{n-1} P_i B_{i,k}(t_0) = \sum_{i=j-k}^{j} P_i B_{i,k}(t_0).$$

b) In the same way, each point P_j influences the curve $S(t)$ for the values of t such that $B_{j,k}(t) \neq 0$, i.e., for $t_j \leq t < t_{j+k+1}$. For instance, if one moves the point P_j, only a small part of the curve $S(t)$ is modified.

2.2.4 Examples

a) Let us look at a quadratic curve ($k = 2$) corresponding to the knot sequence

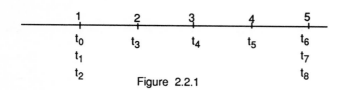

Figure 2.2.1

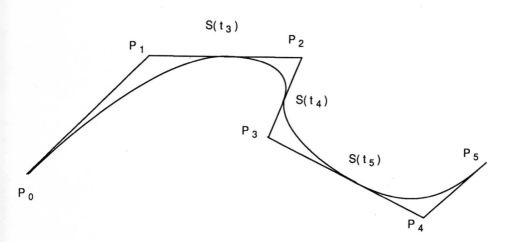

Figure 2.2.2

The reader will check that the curve is tangent to the sides $P_1 P_2$, $P_2 P_3$ and $P_3 P_4$ at their middle, which are the points $S(t_3)$, $S(t_4)$ and $S(t_5)$ respectively (this is special for the case $k = 2$).

b) If we want a *closed* C^1 curve, we may take a uniform sequence of knots (setting for instance $t_i = i$, $i \in \mathbf{Z}$). We then have $B_i(t) = B_{i+r}(t+r) \forall r \in \mathbf{Z}$ (i.e., the B-spline basis elements are integer translates of one of them: see

Figure 2.2.3).

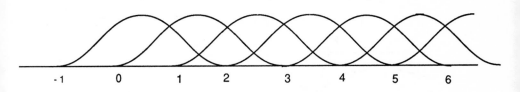

Figure 2.2.3

Let $S(t) = \sum_{i=-\infty}^{+\infty} P_i B_{i,2}(t)$, setting $P_{6+i} = P_i$; we then have $S(t-6) = S(t)$ and the curve is defined on any interval of length 6 (Figure 2.2.4).

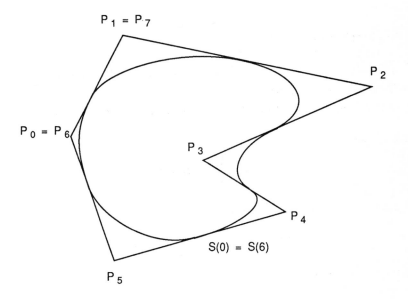

Figure 2.2.4 : Closed spline curve of degree 2

c) Let us draw two Bézier curves of degree 3, for two different shapes of the control polygon:

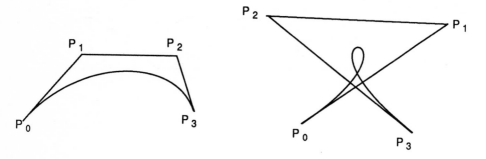

Figure 2.2.5

2.3 ALGORITHMS FOR SPLINE CURVES

a) Evaluation algorithm

This algorithm is also called "De Boor-Cox" in the B-spline case, or "De Casteljau" in the Bézier curve case.

It is a simple adaptation of the algorithm 1.5.1; (see 1.5.2, Remark 2).

2.3.1   Set

$$S(t) = \sum_{i=0}^{n-1} P_i B_{i,k}(t),$$

and assume $t_j \leq t < t_{j+1}$. Then if we set

$$P_i^o = P_i \qquad (j - k \leq i \leq j)$$

and for $0 \leq r \leq k - 1$

$$P_i^{r+1}(t) = \omega_{i,k-r}(t)P_i^r + \big(1 - \omega_{i,k-r}(t)\big)P_{i-1}^r$$
$$= \frac{(t - t_i)P_i^r(t) + (t_{i+k-r} - t)P_{i-1}^r}{t_{i+k-r} - t_i}$$
$$(j - k + r + 1 \leq i \leq j),$$

we have, $P_j^k(t) = S(t).$

Let us illustrate this algorithm with a cubic spline.

We consider the knot sequence

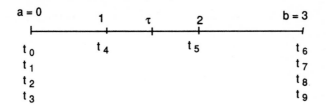

Figure 2.3.1

Let $S(t) = \sum_{i=0}^{5} P_i B_{i,3}(t)$, and let us evaluate $S(\tau)$ for $\tau = 3/2$. The algorithm has a triangular form:

$$
\begin{array}{ccccccc}
P_1 & & P_2 & & P_3 & & P_4 \\
& P_2^1 & & P_3^1 & & P_4^1 & \\
& & P_3^2 & & P_4^2 & & \\
& & & P_4^3 & & &
\end{array}
$$

with $P_4^3 = S(3/2)$ and

$$P_2^1 = 3/4P_2 + 1/4P_1, \qquad P_3^1 = 1/2P_3 + 1/2P_2, \qquad P_4^1 = 1/4P_4 + 3/4P_3,$$
$$P_3^2 = 3/4P_3^1 + 1/4P_2^1, \qquad P_4^2 = 1/4P_4^1 + 3/4P_3^1,$$
$$P_4^3 = 1/2P_4^2 + 1/2P_3^2.$$

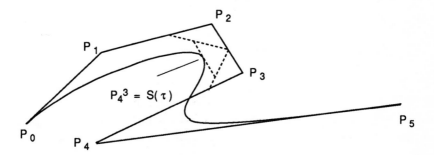

Figure 2.3.2

48   *Ch.2 Spline curves and Bézier curves*

Remarks

1) It is easy to prove that the curve $S(t)$ is *tangent* at P_4^3 to the segment $P_4^2 P_3^2$ (see the Differentiation algorithm below).

2) In the Bézier curve case, the algorithm is simpler, as then the $\omega_{i,k}(t)$'s do not depend on k.

2.3.2 If $B(t) = \sum_{i=0}^{k} P_i B_i^k(t)$ is a Bézier curve (with control points $P_i \in \mathbf{R}^s$), the evaluation algorithm is the following

$$\begin{cases} P_i^o(t) = P_i & (0 \le i \le k-1) \\ P_i^{r+1}(t) = (1-t)P_{i-1}^r + tP_i^r \\ \quad 0 \le r \le k-1, \quad r+1 \le i \le k \\ P_k^k(t) = B(t). \end{cases}$$

b) Differentiation algorithm   The algorithm is explained in 1.5.3.

2.3.4

Let $S(t) = \sum_{i=o}^{n-1} P_i B_{i,k}(t)$ be a spline function, with $t_j \le t < t_{j+1}$. We then have

$$D^r S(t) = \sum_{i=j-k+r}^{j} P_i^r B_{i,k-r}(t)$$

with

$$\begin{cases} P_i^o = P_i & (j-k \le i \le j) \\ P_i^r = \begin{cases} (k-r+1)\dfrac{P_i^{r-1} - P_{i-1}^{r-1}}{t_{i+k-r+1} - t_i} & \text{if } t_i < t_{i+k-r+1} \\ \qquad\qquad (j-k+r \le i \le j) \\ 0 & \text{otherwise.} \end{cases} \end{cases}$$

To evaluate the derivative $D^r S(t)$ at a point t, one may then use the above evaluation algorithm to compute $B_{i,k-r}(t)$; this allows us for instance to find the values of the successive derivatives at a knot t_i, and so to find the polynomial expression of the curve $S(t)$ between t_i and t_{i+1}, with Taylor's formula.

Let us look at the case of Bézier curves: let $B(t) = \sum_{i=0}^{k} P_i B_i^k(t)$ be a Bézier curve of degree k.

Set

$$\Delta P_i = P_{i+1} - P_i,$$
$$\Delta^2 P_i = \Delta(\Delta P_i) = P_{i+2} - 2P_{i+1} - P_i,$$
$$\Delta^k P_i = \Delta(\Delta^{k-1} P_i) = \sum_{j=0}^{k} (-1)^{k-j} \binom{k}{j} P_{i+j}.$$

2.3.5 Lemma *We have*

$$D^r B(t) = \frac{k!}{(k-r)!} \sum_{i=0}^{k-r} B_i^{k-r} \Delta^r P_i.$$

Proof Easy by induction on r, using 2.1.3, e), or the formula for B-splines studied above. ∎

If we now want to evaluate $D^r B$ at t, we may use the De Casteljau's algorithm 2.3.2.

Note that this algorithm commutes with the Δ operation, as the latter is linear $(\Delta(\lambda P_{i-1} + \mu P_i) = \lambda \Delta P_{i-1} + \mu \Delta P_i)$.

The evaluation algorithm of B and its derivatives $D^r B$ at point t becomes then:

2.3.6

a) Evaluate $B(t)$ with the algorithm 2.3.2, which gives a triangular array $T^{(0)}$ of points P_i^r $(0 \le r \le k, r \le i \le k)$.

b) Pass from array $T^{(l)}$ to array $T^{(l+1)}$ by difference (operation Δ).

c) $D^r B(t)$ is given by the last element of the array $T^{(r)}$.

Let us treat as an example the case $k = 3$; the algorithm 2.3.2 gives the following array

$$
\begin{array}{ccccccc}
P_0 & & P_1 & & P_2 & & P_3 \\
 & P_1^1 & & P_2^1 & & P_3^1 & \\
 & & P_2^2 & & P_3^2 & & \\
 & & & P_3^3 & & &
\end{array}
$$

with $P_i^{r+1} = (1-t)P_{i-1}^r + tP_i^r$, and, by difference, the following arrays

$$
\begin{array}{cccccc}
P_1 - P_0 & & P_2 - P_1 & & P_3 - P_2 \\
 & P_2^1 - P_1^1 & & P_3^1 - P_2^1 & \\
 & & P_3^2 - P_2^2 & &
\end{array}
$$

$T^{(1)}$

$T^{(2)}$ $\qquad P_2 - 2P_1 + P_0 \qquad\qquad\qquad P_3 - 2P_2 + P_1$

$$P_3^1 - 2P_2^1 + P_1^1$$

$T^{(3)}$ $\qquad\qquad\qquad\qquad P_3 - 3P_2 + 3P_1 - P_0.$

We then get

$(2.3.7,1)$
$$\begin{cases} B(t) = P_3^3 \\ \dfrac{1}{3}B'(t) = P_3^2 - P_2^2 \\ \dfrac{1}{6}B''(t) = P_3^1 - 2P_2^1 + P_1^1 \\ \dfrac{1}{6}B'''(t) = P_3 - 3P_2 + 3P_1 - P_0. \end{cases}$$

For instance, if $t = 0$, one has $P_i^r = P_{i-r} \quad \forall\ r$, and therefore

$(2.3.7,2)$
$$\begin{cases} B(0) = P_0 \\ \dfrac{1}{k}B'(O) = P_1 - P_0 \\ \dfrac{1}{k(k-1)}B''(0) = P_1 - 2P_1 + P_0 \\ \dfrac{1}{k(k-1)(k-2)}B'''(0) = P_3 - 3P_2 + 3P_1 - P_0 \end{cases}$$

(these formulas are valid for any Bézier curve of degree $k > 2$).

If now $t = 1$, one has $P_i^r = P_i$, which gives, for B of degree 3

$(2.3.7,3)$
$$\begin{cases} B(1) = P_3 \\ \dfrac{1}{3}B'(1) = P_3 - P_2 \\ \dfrac{1}{6}B''(1) = P_3 - 2P_2 + P_1 \\ \dfrac{1}{6}B'''(1) = P_3 - 3P_2 + 3P_1 + P_0, \end{cases}$$

which lets us give conditions of junction between two Bézier curves.

c) Knot insertion algorithm

The algorithm is described in 1.5.4. Recall that if

$$S(t) = \sum_{i=0}^{n-1} P_i B_{i,k}(t)$$

and if we add a new knot $\hat{t}$, we have $S(t) = \sum_{i=0}^{n} \hat{P}_i \hat{B}_{i,k}(t)$, with:

$$(2.3.8) \qquad \hat{P}_i \;=\; \begin{cases} P_i & \text{if } t_{i+k} \leq \hat{t}, \\ \omega_{i,k}(\hat{t})P_i + \left(1 - \omega_{i,k}(\hat{t})\right)P_{i-1} & \text{if } t_i < \hat{t} < t_{i+k}, \\ P_{i-1} & \text{if } \hat{t} \leq t_i. \end{cases}$$

We see that these formulas are the same as those in first step of the evaluation algorithm 2.3.1 (if for instance we have $t_j < \hat{t} < t_{j+1}$, the condition $t_i < \hat{t} < t_{i+k}$ is equivalent to $j - k + 1 \leq i \leq j$, which corresponds to the statement 2.3.1).

We see again that, if we insert successively k knots at the same place τ (with $t_j < \tau < t_{j+1}$), and if at the rth time we get the points P_i^r $(P_i^0 = P_i)$, then $S(\tau) = P_j^k$.

Let us look at the situation after the insertion of k knots at τ, and assume $t_j < \tau < t_{j+1}$. If we set $(t_i')_{0 \leq i \leq n+2k}$ for the new knot sequence, the support of the B-spline $B_{j,k,t'}$ is $[t_j, t_{j+1}]$ (as then $t_{j+1} = t_{j+k+1}'$), and one has $B_{j,k,t'}(\tau) = 1$ (see 1.3.4, e) and Figure 2.3.3).

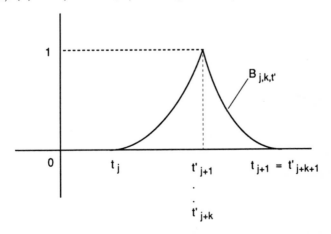

Figure 2.3.3

We may then write, for a new sequence (P'_i) of control points

$$S(t) = \sum_{i=0}^{n+k-1} P'_i B_{i,k,t'}(t),$$

with $S(\tau) = P'_j$.

Let us add now a $(k+1)^{th}$ knot $\hat{t}$ equal to τ. Formulas 2.3.8 give then

$$\begin{cases} \hat{P}_i = P'_i & \text{for } i \leq j \text{ and} \\ \hat{P}_i = P'_{i-1} & \text{for } i > j. \end{cases}$$

In other words, the point P'_j has been *doubled* and the B-spline $B_{j,k,t'}$ "divided in two" (one half $\hat{B}_{j,k}$ with support $[t_j, \tau]$ and the other $\hat{B}_{j+1,k}$ with support $[\tau, t_{j+1}] = [\tau, t'_{j+k+1}]$). The supports of all the B-splines $\hat{B}_{i,k}$ are then contained either in $[a, \tau]$, or in $[\tau, b]$, and the spline curve $S(t)$ has split in two spline curves (of the same degree k), one defined on $[a, \tau]$ and the other on $[\tau, b]$. These two curves are defined in the following way (with the notation of 2.3.1)

$$(2.3.8, 1) \qquad\qquad S_1(t) = \sum_{i=0}^{j} P_i^k \hat{B}_{i,k}(t) \qquad \text{for } a \leq t \leq \tau,$$

and

$$(2.3.8, 2) \qquad\qquad S_2(t) = \sum_{i=j+1}^{n+k-1} P_i^k \hat{B}_{i,k}(t) \qquad \text{for } \tau \leq t \leq b.$$

2.3.9 Remark   One has in fact $\hat{B}_{j,k}(\tau) = 0$ and $\hat{B}_{j+1,k}(\tau) = 1$, as the B-splines are always right continuous. In the formula (2.3.8, 1) above, one generally sets $\hat{B}_{j,k}(\tau) = 1$ so that $S_1(t)$ be continuous at τ (giving $S(t) = S_1(t) + S_2(t)$ except for $t = \tau$).

This preceding fact allows us to subdivide the spline curves; we will begin by studying in detail the case of Bézier curves, for which the algorithm is especially simple.

d) Subdivision algorithm for a Bézier curve.

Set for instance $\tau = 1/2$. If $B(t) = \sum_{i=0}^{k} P_i B_i^k(t)$ is a Bézier curve of degree k, the algorithm for the evalution of $B(1/2)$ is then (see 2.3.2)

$$
\begin{cases}
P_i^o = P_i & (0 \le i \le k) \\
P_i^{j+1} = 1/2(P_{i-1}^j + P_i^j) & (j+1 \le i \le k) \\
P_k^k = B(1/2)
\end{cases}
$$

After duplicating once the point P_k^k we see, applying what has been done for the splines in c) above, that the curve has split in two curves:

$$
B_1(t) = \sum_{i=0}^{k} P_i^i \widehat{B}_i^k(t) \quad \text{for} \quad 0 \le t \le 1/2,
$$

with $\widehat{B}_i^k(t) = B_i^k(2t)$ (see 2.1.5) and

$$
B_2(t) = \sum_{i=0}^{k} P_i^i \widehat{B}_i^{\prime k}(t) \quad \text{for} \quad 1/2 \le t \le 1,
$$

with this time $\widehat{B}_i^{\prime k}(t) = B_i^k(2t - 1)$ (see 2.1.5).

Example ($k = 3$)   The algorithm may be represented with a triangular array

$$
\begin{array}{ccccccc}
P_0 & & P_1 & & P_2 & & P_3 \\
 & P_1^1 & & P_2^1 & & P_3^1 & \\
 & & P_2^2 & & P_3^2 & & \\
 & & & P_3^3 & & &
\end{array}
$$

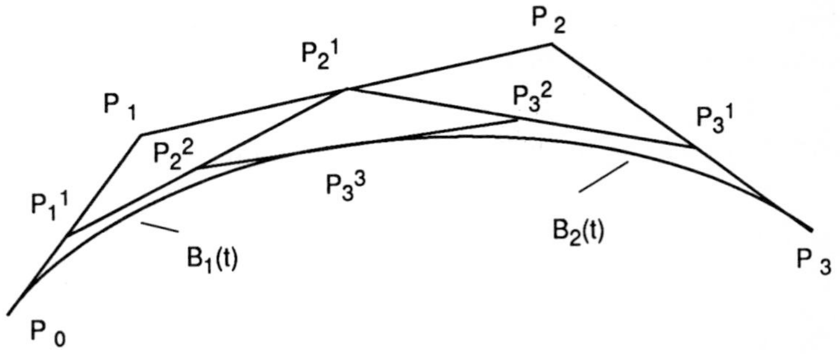

Figure 2.3.4

We may now iterate the process, applying the same algorithm to $B_1(t)$ for $t = 1/4$, and to $B_2(t)$ for $t = 3/4$. Carrying on with this process, one obtains at the nth step 2^n Bézier curves whose union is the initial curve, and 2^n control polygons Π_n^i whose total number of vertices is $2^n k + 1$.

Among those vertices, there are $2^n + 1$ points of the initial curve: the points corresponding to $t = p/2^n$ (p integer, $0 \leq p \leq 2^n$).

2.3.10 Proposition   *The sequence of polygons Π_n^k converges uniformly towards the Bézier curve $B(t)$, with speed $\lambda/2^n$ in the case where $\tau = 1/2$; λ is a constant independent of n.*

Proof   Assume $\tau = 1/2$; it is clear that if M designates the length of the greatest side of the control polygon of a Bézier curve $B(t)$ ($M = \text{Sup} \, |P_i P_{i+1}| \, 0 \leq i \leq k$), the distance of this polygon to the curve is less than kM, and that the length of the greatest side of one of the control polygons of one of the 2^n Bézier curves obtained at the nth step has length $\leq M/2^n$, which proves 2.3.10.

∎

2.3.11 Remarks   The subdivision algorithm gives the vertices of the polygon Π_n^i; after a certain step, and considering the screen resolution, the polygon Π_n^i is "indistinguishable" from the curve. This allows a fast tracing (as the convergence is exponential) of a Bézier curve $B(t)$.

e) "Oslo" algorithm for a spline curve

Let $B(t) = \sum_{i=0}^{n-1} P_i B_{i,k}(t)$ be a spline curve corresponding to the knot sequence $t = (t_0, \ldots, t_{n+k})$, let $\bar{t} = (\bar{t}_1, \ldots, \bar{t}_{m+k})$ a finer knot sequence (i.e., such that $t \subset \bar{t}$, and such that, as always $\bar{t}_i < \bar{t}_{i+k+1} \, \forall i$), and let $\overline{B}_{i,k}$ be the corresponding B-splines.

Each $B_{i,k}$ is, in an unique way, expressed in function of the $\overline{B}_{j,k}$'s:

$$B_{i,k}(t) = \sum_j \alpha_{i,k}(j) \overline{B}_{j,k}(t),$$

which implies, if we set $B(t) = \sum_{j=0}^{m-1} Q_j \overline{B}_{j,k}(t)$, the relation $Q_j = \sum_i \alpha_{i,k}(j) P_i$ which computes the points Q_j of the new control polygon of the curve $B(t)$.

For the computation of the points Q_j, we may
- either use as many times as necessary the knot insertion algorithm described in 2.3.8.
- or use the "Oslo" algorithm described below. These two proceses are equivalent, and have similar complexities. Consequently, we will only describe the Oslo algorithm, without any justification.

Let us define the numbers $\alpha_{i,l}(j)$ for $l < k$ in the following way

$$B_{i,l}(t) = \sum_j \alpha_{i,l}(j)\overline{B}_{j,l}(t)$$

with by definition $\alpha_{i,l}(j) = 0$ if $\overline{B}_{j,l} \equiv 0$ i.e., if $\bar{t}_j = \bar{t}_{j+l+1}$.

2.3.12 Proposition

One has

$$\alpha_{i,0}(j) = \begin{cases} 1 & \text{if } t_i \leq \bar{t}_j < t_{i+1} \\ 0 & \text{if not} \end{cases}$$

and for k ¿ 0

$$\alpha_{i,k}(j) = \omega_{i,k}(\bar{t}_{j+k})\alpha_{i,k-1}(j) + (1 - \omega_{i+1,k}(\bar{t}_{j+k}))\alpha_{i+1,k-1}(j)$$
$$= \frac{\bar{t}_{j+k} - t_i}{t_{i+k} - t_i}\alpha_{i,k-1}(j) + \frac{t_{i+k+1} - \bar{t}_{j+k}}{t_{i+k+1} - t_{i+1}}\alpha_{i+1,k-1}(j) \quad \text{if } \bar{t}_j < \bar{t}_{j+k}$$

when these expressions are defined, and

$$\alpha_{i,k}(j) = \frac{\bar{t}_{j+k} - t_i}{t_{i+k} - t_i}\alpha_{i,k-1}(j+1) + \frac{t_{i+k+1} - \bar{t}_{j+k+1}}{t_{i+k+1} - t_{i+1}}\alpha_{i+1,k-1}(j+1)$$

if $\bar{t}_j = \bar{t}_{j+k}$.

Note that if $\bar{t}_j = \bar{t}_{j+k}$, we have $\alpha_{i,k-1}(j) = 0 \quad \forall j$.

2.3.13 Remark We immediately see by induction that $\alpha_{i,k}(j) = 0$ if

$$\begin{cases} \bar{t}_j < t_i & \text{or} \\ t_{i+k+1} \leq \bar{t}_j. \end{cases}$$

The Oslo algorithm allows us therefore to compute the Q_j's in terms of the P_i's. One may also compute them directly with the following algorithm

2.3.14   Assume $t_s \le \bar{t}_j < t_{s+1}$ and $\bar{t}_j < \bar{t}_{j+1}$; then set

a)
$$P_{i,j}^0 = P_i \quad \text{for} \quad s - k \le i \le s$$

b)
$$P_{i,j}^{r+1} = \frac{(\bar{t}_{j+k-r} - t_i)P_{i,j}^r + (t_{i+k-r} - \bar{t}_{j+k-r})P_{i-1,j}^r}{t_{i+k-r} - t_i}$$

for $0 \le r \le k-1$ and $s - k + r + 1 \le i \le s$.

c)
$$\text{One has then} \quad P_{s,j}^k = Q_j$$

Proof   We have by definition

$$Q_j = \sum_j \alpha_{i,k}(j)P_i = \sum_{i=s-k}^{s} \alpha_{i,k}(j)P_{i,j}^0$$

$(\alpha_{i,k}(j) = 0$ if $i \notin [s-k, s]$ by Remark 2.3.13), therefore

$$Q_j = \sum_{i=s-k+1}^{s} \alpha_{i,k-1}(j)P_{i,j}^1$$

applying 2.3.12 and the definition of $P_{i,j}^1$, and the result follows by induction.
∎

This algorithm needs the calculation of $\frac{k(k+1)}{2}$ linear combinations for each Q_j: the complexity is the same as if we apply k times the knot insertion algorithm 2.3.8.

2.3.15 Example

Let us look at two knot sequences:

$$\tau = \big(t_0 = t_1 = t_2 = t_3 < t_4 < t_5 < \ldots\big)$$

$$\bar{\tau} = \big(\bar{t}_0 = \bar{t}_1 = \bar{t}_2 = \bar{t}_3 < \bar{t}_4 < \bar{t}_5 < \ldots\big)$$

with $t_i = \bar{t}_i$ for $0 \le i \le 3$, $\bar{t}_4 = 1/2(t_3 + t_4)$, $\bar{t}_5 = t_4$, $\bar{t}_6 = 1/2(t_4 + t_5)$, $\bar{t}_7 = t_5$, etc … (one adds a new knot at the middle of each interval $[t_i, t_{i+1}]$, $(i \ge 3)$), and let us take cubic splines ($k = 3$).

1) We apply the algorithm 2.3.14 above

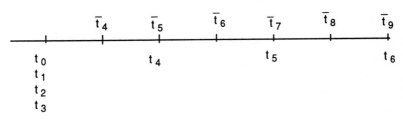

Figure 2.3.5

One assume that the t_i's are integers, i.e., for instance that $t_0 = t_1 = t_2 = t_3 = 3$, $t_n = n$ $(n \geq 3)$.

The algorithm then gives

$$
\begin{cases}
Q_0 = P_0, \qquad Q_1 = 1/2(P_0 + P_1) \\
Q_2 = \dfrac{3}{4}P_1 + \dfrac{1}{4}P_2 \\
Q_3 = \dfrac{3}{16}P_1 + \dfrac{11}{16}P_2 + \dfrac{2}{16}P_3.
\end{cases}
$$

After the computation of Q_4, the construction is periodic and gives:

$$
\begin{cases}
Q_{2i} = \dfrac{1}{2}(P_i + P_{i+1}) \\
Q_{2i+1} = \dfrac{1}{8}(P_i + 6P_{i+1} + P_{i+2}),
\end{cases}
$$

see Figure 2.3.6 below.

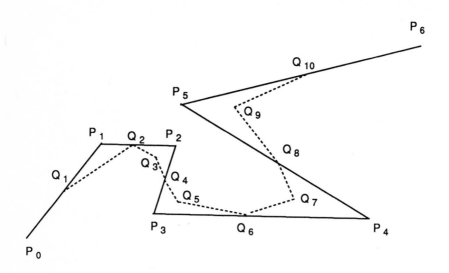

Figure 2.3.6

As an illustration, we evaluate Q_3: we have $k = 4$, $s = 3$, and so $P_{i,3}^0 = P_i$ for $0 \leq i \leq 3$; set, for simplification, $P_{i,3}^r = P_i^r$ $(1 \leq r \leq 3)$;

$r = 0$ P_0 P_1 P_2 P_3

$r = 1$ $P_1^1 = \frac{3}{2}P_1 - \frac{1}{2}P_0$ $P_2^1 = \frac{3}{4}P_2 + \frac{1}{4}P_1$ $P_3^1 = \frac{1}{2}(P_3 + P_2)$.

$r = 2$ $P_2^2 = P_2^1$ $P_3^2 = \frac{1}{2}(P_3^1 + P_2^1)$

$r = 3$ $P_3^3 = \frac{1}{2}(P_3^2 + P_2^2)$,

and finally we get $Q_3 = P_3^3 = \frac{3}{16}P_1 + \frac{11}{16}P_2 + \frac{2}{16}P_3$.

2) We now apply successively the knot insertion algorithm.

Insertion of $\bar{t}_4$

We have

$$P_0^1 = P_0,$$

$$P_1^1 = \frac{1}{2}(P_0 + P_1),$$

$$P_2^1 = \frac{3}{4}P_1 + \frac{1}{4}P_2,$$

$$P_3^1 = \frac{1}{6}P_3 + \frac{5}{6}P_2,$$

$$P_4^1 = P_3,$$

$$P_5^1 = P_4,$$

etc.

Insertion of $\bar{t}_6$:

$$Q_0 = P_0^2 = P_0^1,$$

$$Q_1 = P_1^2 = P_1^1,$$

$$Q_2 = P_2^2 = P_2^1,$$

$$Q_3 = P_3^2 = \frac{2}{5}P_1^1 + \frac{3}{5}P_3^1 = \frac{1}{2}(P_2 + P_3),$$

$$P_5^2 = \frac{1}{6}P_5^1 + \frac{5}{6}P_4^1 = \frac{1}{6}P_4 + \frac{5}{6}P_3,$$

$$P_6^2 = P_5^1, \text{ etc.}$$

There are three new points to compute at each step, and the complexity is similar.

2.3.16 Remark   If we apply n times the Oslo algorithm in the case of Example 2.3.15, i.e., adding at each step a new knot in the middle of the interval $[t_i, t_{i+1}])$, and if Π_n is the control polygon obtained at the nth step, one can prove that the polygon Π_n converges towards the spline curve $B(t)$ in $O(\frac{1}{2^n})$ (see 2.3.10).

f) Subdivision algorithm for uniform splines

Let $B_k(X)$ be the spline of degree k, built with integer knots, and with support $[0, \ldots, k+1]$. Its translates $B_k(X - j)$ satisfy $\sum_{j \in \mathbf{Z}} B_k(X - j) = 1$, as results from 1.3.4 applied to a sufficiently large interval $[a, b]$ containing X. The $B_k(X - j)$'s are by definition the *uniform B-splines* corresponding to the knot sequence $\tau = \mathbf{Z}$.

Let now $\tau_p = \{j/p, \ j \in \mathbf{Z}\}$, $p \geq 2$ be a finer uniform knot sequence: the corresponding B-splines of degree k are the $B_k(pX - j)$'s, and we may expand $B_k(X)$ with respect to the basis $B_k(pX - j)$

$$B_k(X) = \sum_j a_p^k(j/p)B_k(pX - j).$$

2.3.17 Remark   The support of $B_k(X)$ is $[0, k + 1]$ and that of $B_k(pX - j), [j/p, \frac{j+k+1}{p}]$; we then have

$$B_k(X) = \sum_{j=-k}^{p(k+1)-1} a_p^k(j/p)B_k(pX - j).$$

But $B_k(X) \mid [\frac{-k}{p}, 0] \equiv 0$; note that the functions $B_k(pX - j) \mid [\frac{-k}{p}, 0]$ $(-k \leq j \leq -1)$ are *linearly independent*, as results from Theorem 1.4.3 applied to the interval $[a, b] = [\frac{-k}{p}, 0]$ and to the knot sequence (j/p) $(-2k \leq j \leq k)$. We then have $a_p^k(j/p) = 0$ $(-k \leq j \leq -1)$.

In the same way, we have $B_k(X)|[k + 1, k + 1 + k/p] \equiv 0$, which implies by the same argument that $a_p^k(j/p) = 0$ for $p(k + 1) - k \leq j \leq p(k + 1) - 1$. So finally, we may write

$$B_k(X) = \sum_{j=0}^{(k+1)(p-1)} a_p^k(j/p)B_k(pX - j),$$

which we write also as

$$B_k(X) = \sum_j a_p^k(j/p)B_k(pX - j),$$

setting $a_p^k(j/p) = 0$ for $j \notin [0, (k + 1)(p - 1)]$.

Let now $B(t) = \sum_{i \in \mathbf{Z}} P_i B_k(t - i)$ be a spline curve with control polygon (P_i) ;$B(t)$ is well defined, as the sum is finite for each t. It is often convenient to consider an infinite sequence of P_i's; one may return to the notation of Section 1, looking at functions on $[a, b] = [k, n]$ for instance, and to the knots $(t_i)_{0 \leq i \leq n+k}$.

With the above notation and conventions, (see Remark 2.3.17), we have

$$B(t) = \sum_{j \in \mathbf{Z}} Q_p^k(j/p)B_k(pt - j),$$

where the new control polygon $Q_p^k(j/p)$ satisfies the following relations

$$Q_p^k(j/p) = \sum_{i \in \mathbf{Z}} a_p^k(j/p - i)P_i.$$

The subdivision algorithm then comes from the following Lemma

2.3.18 Lemma   *The numbers $a_p^k(j/p)$ are the coefficients of the devlop-ment of the rational fraction $\frac{1}{p^k}\left(\frac{Z^p-1}{Z-1}\right)^{k+1}$; in other words, one has*

$$\frac{1}{p^k}\left(\frac{Z^p-1}{Z-1}\right)^{k+1} = \sum_{j=0}^{(k+1)(p-1)} a_p^k(j/p)Z^j.$$

Proof   This is by induction on k, the case $k = 0$ being trivial ($a_p^0(j/p) = 1$ for $0 \leq j \leq p-1$) as $B_0(t)$ is the characteristic function of the interval $[0, 1[$, and $B_0(pt - j)$ that of the interval $[\frac{j}{p}, \frac{j+1}{p}[$).

The general case is a simple application of the Definition of B-splines 1.3.2. We have

$$B_k(t) = \frac{1}{k}\Big(tB_{k-1}(t) + (k+1-t)B_{k-1}(t-1)\Big)$$

by 1.3.2, therefore

$$B_k(pt - j) = \frac{1}{k}\Big[(pt-j)B_{k-1}(pt-j) + (k+1+j-pt)B_{k-1}(pt-j-1)\Big].$$

One deduces that

a) $$B_k(t) = \frac{1}{k}\sum_{j=0}^{(k+1)(p-1)} a_p^k(j/p)\Big[(pt-j)B_{k-1}(pt-j)$$

$$+(k+j+1-pt)B_{k-1}(pt-j-1)\Big]$$

using the formula

$$B_k(t) = \sum_{j=0}^{(k+1)(p-1)} a_p^k(j/p)B_k(pt-j).$$

b) Replacing $B_{k-1}(t)$ and $B_{k-1}(t-1)$ by their development (in the formula $B_k(t) = \frac{1}{k}\left(tB_{k-1}(t) + (k+1-t)B_{k-1}(t-1)\right)$), gives also

$$B_k(t) = \frac{1}{k}\left[t \sum_{j=0}^{k(p-1)} a_p^{k-1}(j/p)B_{k-1}(pt-j)\right.$$

$$\left. + (k+1-t) \sum_{j=0}^{k(p-1)} a_p^{k-1}(j/p)B_{k-1}(pt-j-p)\right].$$

We obtain thus the relation:

$$\sum_{j=0}^{(k+1)(p-1)} B_{k-1}(pt-j)(tA_j + B_j) = 0$$

with

$$(2.3.18,1) \quad A_j = p\left[a_p^k(j/p) - a_p^k\left(\frac{j-1}{p}\right)\right] - a_p^{k-1}(j/p) + a_p^{k-1}\left(\frac{j-p}{p}\right),$$

$$(2.3.18,2) \qquad B_j = (k+j)a_p^k\left(\frac{j-1}{p}\right) - ja_p^k(j/p) - (k+1)a_p^{k-1}\left(\frac{j-p}{p}\right)$$

for $0 \le j \le (k+1)(p-1)$, always with the convention that $a_p^r(j/p) = 0$ for $j < 0$ and $a_p^r(j/p) = 0$ for $j > (r+1)(p-1)$. By the induction hypothesis, one has the relation

$$\frac{1}{p^{k-1}}\left(\frac{Z^p - 1}{Z - 1}\right)^k = \sum_{j=0}^{k(p-1)} a_p^{k-1}(j/p)Z^j.$$

Let us define the numbers $b_p^k(j/p)$ $(j \in \mathbf{Z})$ by:

$$b_p^k(j/p) = 0 \text{ for } j < 0 \text{ or } j > (k+1)(p-1),$$

and by the formula

$$\frac{1}{p^k}\left(\frac{Z^p - 1}{Z - 1}\right)^{k+1} = \sum_{j=0}^{(k+1)(p-1)} b_p^k(j/p)Z^j.$$

We obtain, using the induction hypothesis, the relation

$$p(Z-1)\sum_{j=0}^{(k+1)(p-1)} b_p^k(j/p)Z^j = (Z^p-1)\sum_{j=0}^{k(p-1)} a_p^{k-1}(j/p)Z^j,$$

whence, equating the coefficients of Z^j

$$(2.3.18,3) \quad p\left[b_p^k(j/p) - b_p^k\left(\frac{j-1}{p}\right)\right] = a_p^{k-1}(j/p) - a_p^{k-1}\left(\frac{j-p}{p}\right) \quad (j \in \mathbf{Z}).$$

In the same way, differentiating in Z the equation

$$\frac{1}{p^k}(Z^p-1)^{k+1} = (Z-1)^{k+1}\sum_{j=0}^{(k+1)(p-1)} b_p^k(j)Z^j,$$

using induction hypothesis, and equating the coefficients of Z^{j-1}, we obtain

$$(2.3.18,4) \qquad (k+1)a_p^{k-1}\left(\frac{j-p}{p}\right) = (k+j)b_p^k\left(\frac{j-1}{p}\right) - jb_p^k(j/p).$$

Now, let us set

$$C_k(t) = \sum b_p^k(j/p)B_k(pt-j);$$

we have as above

$$C_k(t) = \frac{1}{k}\sum b_p^k(j/p)\Big[(pt-j)B_{k-1}(pt-j) + (k+j+1)B_{k-1}(pt-j-1)\Big],$$

therefore

$$B_k(t) - C_k(t) =$$
$$\frac{1}{k}\sum B_{k-1}(pt-j)[pt\Big((a_p^k(j/p) - a_p^k((j-1)/p)\Big) - pt\big(b_p^k(j/p) - b_p^k((j-1)/p)\big)$$

$$+\Big((k+j)a_p^k((j-1)/p) - ja_p^k(j/p)\Big) - \Big((k+j)b_p^k((j-1)/p) - jb_p^k(j/p)\Big)$$

$$= \frac{1}{k}\sum_{j=0}^{(k+1)(p-1)} B_{k-1}(pt-j)TA_j + B_j) = 0,$$

using (2.3.18, 3) and (2.3.18, 4), and the definitions (2.3.18, 1) and (2.3.18, 2) of A_i and B_i, which proves that $B_P^k(j/p) = a_p^k(j/p)$.

2.3.19 Corollary   ("Subdivision algorithm")

With the above notation, the following induction formula holds for the control polygons

$$Q_p^{k+1}(j/p) = \frac{1}{p}\sum_{l=0}^{p-1} Q_p^k(j/p - l/p) \quad (j \in \mathbf{Z}).$$

Proof   Recall that $Q_p^k(j/p) = \sum_{i\in\mathbf{Z}} a_p^k(j/p - i)P_i$.

Lemma 2.3.18 gives the relation

$$\frac{1}{p}\left(\frac{Z^p - 1}{Z - 1}\right)\left(\sum_{j\in\mathbf{Z}} a_p^k(j/p)Z^j\right) = \sum_{j\in\mathbf{Z}} a_p^{k+1}(j/p)Z^j,$$

whence $a_p^{k+1}(j/p) = \frac{1}{p}\sum_{l=0}^{p-1} a_p^k(\frac{j-l}{p})$, which proves the Corollary.

∎

In particular, we obtain for $p = 2$

$$\begin{cases} Q_2^{k+1}(j) = \frac{1}{2}(Q_2^k(j) + Q_2^k(j - \frac{1}{2})) \\ Q_2^{k+1}(j + \frac{1}{2}) = \frac{1}{2}(Q_2^k(j + \frac{1}{2}) + Q_2^k(j)), \end{cases}$$

which allows a very easy construction of the control polygon $Q_p^k(j/p)$, starting from the relations $Q_p^0(j) = \cdots = Q_p^0(j + \frac{p-1}{p}) = P_j$ $(j \in \mathbf{Z})$, which come from the fact that $a_p^0(j/p) = 1$ for $0 \le j \le p - 1$, and 0 otherwise.

2.3.20 Example   Let us look at the following closed polygon

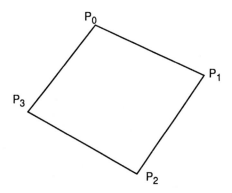

Figure 2.3.7

We set $P_j = P_{j+4}$, and $B(t) = \sum_{j \in \mathbf{Z}} P_j B_3(t - j)$.

We obtain then in three steps, for the same periodic spline, the control polygon associated to a twice as fine subdivision

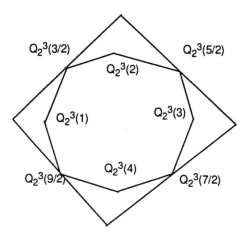

Figure 2.3.8

As for the Oslo algorithm, the convergence of the polygons (Q_p^k) towards the curve $B(t)$ is in $\frac{1}{p^2}$; it is so very fast (i.e., exponential) if we choose a sequence of sudivisions such that each new one be twice as fine as the preceding one.

3

Interpolation and complements

3.1 INTERPOLATION

Let $P_0, \ldots, P_{n-1}$ be n points in $\mathbf{R}^s$; we look for a spline curve $B(t)$ of degree k passing through the points P_i. For that, let us take a knot sequence $(t_i)_{0 \leq i \leq n+k}$ (as always, we assume that $t_i < t_{i+k+1}, 0 \leq i \leq n-1$). We then try to find control points Q_j and points $u_i \in \mathbf{R}$ $(u_i \leq u_{i+1})$ such that the curve $B(t) = \sum_{j=0}^{n-1} B_{j,k}(t)Q_j$ satisfy $B(u_i) = P_i$ $(0 \leq i \leq n-1)$; we then have to solve the linear system

$$\sum_{j=0}^{n-1} B_{j,k}(u_i)Q_j = P_i \qquad (0 \leq i \leq n-1).$$

3.1.1 Theorem *The matrix $N = \left(B_{j,k}(u_i) \right)$ is invertible if and only if all the diagonal elements are non zero, i.e., if $B_{j,k}(u_j) \neq 0$ $(0 \leq j \leq n-1)$, or else $t_i < u_i < t_{i+k+1}$ $(0 \leq i \leq n-1)$.*

3.1.2 Remarks

1) One may possibly have $u_i = t_i$ and $Q_i = P_i$ when t_i is a knot of order $k+1$ $(t_i = t_{i+1} = \cdots = t_{i+k} < t_{i+k+1}$: see 1.3.4).

2) One generally chooses $u_i = t_i^* = \frac{t_{i+1} + \cdots + t_{i+k}}{k}$ (see Chapter 1, Section 6.1), which implies that the condition of Theorem 3.1.1 is fulfilled, except when t_{i+1} is a knot of order $k+1$ $(t_i < t_{i+1} = \cdots = t_{i+k+1})$.

Moreover, the choice $u_i = t_i^*$ implies that the spline curve will have less oscillation than the polygon (P_i) (see 1.5.6 and 1.6.4), i.e., that this oscillation will be "minimal" among all the interpolating curves.

3) In some cases, the values of the parameters u_i (such that $B(u_i) = P_i$) are a priori fixed, but we have to choose the values of t_j; we then generally set

$$\begin{cases} t_0 = \cdots = t_k = u_0 \\ t_{i+k} = \dfrac{u_i + \cdots + u_{i+k-1}}{k} \quad (1 \le i \le n-k-1) \\ t_n = \cdots = t_{n+k} = u_{n-1}. \end{cases}$$

Proof of 3.1.1

Proof of necessity.

Assume that there exists i such that $B_i(u_i) = 0$ (with $0 \le i \le n-1$); assume for instance that $u_i \le t_i$ (and $t_i < t_{i+k}$ if $u_i = t_i$, see 1.3.4).

Then each of the $(i+1)$ first rows of the matrix N contains at most i non zero terms (namely $B_0(u_r), \ldots, B_{i-1}(u_r)$ for the rth row $(r \le i)$, as $B_j(u_r) = 0$ for $j \ge i$), and then these $(i+1)$ rows are dependent, which implies that the matrix N is singular.

Proof of sufficiency.

Let

$$\sum_{j=0}^{n-1} \lambda_j B_j(u_i) = 0 \quad (0 \le i \le n-1)$$

be a relation between the columns of N. Set $B(t) = \sum_{j=0}^{n-1} \lambda_j B_j(t)$: we have then by hypothesis $B(u_i) = 0$ $(0 \le i \le n-1)$. We have to prove that, with the hypothesis $B_i(u_i) \neq 0$ $(0 \le i \le n-1)$, this implies $\lambda_j = 0$ $(0 \le j \le n-1)$.

3.1.3 Lemma a) *Assume $i \ge k$ and $t_i < t_{i+1}$; the B-splines $B_{i-k}(t), \ldots, B_i(t)$ restricted to the interval $[t_i, t_{i+1}[$ are linearly independent.*

b) *If $0 \le i < k$, and $t_i < t_{i+1}$, then the B-splines $B_0(t), \ldots, B_i(t)$ restricted to the interval $[t_i, t_{i+1}]$ are linearly independent.*

Proof a) One may use directly the Theorem 1.4.3 (setting $[a, b] = [t_i, t_{i+1}]$), or use the following argument: the space of splines (i.e., the space $\mathcal{P}_{k,t}$, see 1.4.2) restricted to the interval $[t_i, t_{i+1}[$ can be identified with the space of polynomials of degree $\leq k$. As the space is of dimension $k + 1$ and is spanned by the $k + 1$ functions $B_{i-k}(t), \dots, B_i(t)$ (1.4.4), these $k + 1$ functions are linearly independent.

b) results from a), as is seen by adding knots $t_{i-k} \leq \dots \leq t_{-1} \leq t_0$ and looking at the corresponding B-splines.

■

We deduce from this lemma that if $B(t) \equiv 0$ for $t \in [t_i, t_{i+1}[$ ($t_i < t_{i+1}$), one has $\lambda_{i-k} = \dots = \lambda_i = 0$ (the λ's with indices < 0 are equal to zero by convention). Moreover if, for all the intervals $[t_j, t_{j+k+1}[$, there exists i ($j \leq i \leq j+k+1$, $t_i < t_{i+1}$) such that $B(t) \big| [t_i, t_{i+1}[\; \equiv 0$, then $B(t) \equiv 0$.

We then may assume that there exists an $I = [t_r, t_s[$ such that:

a) One does not have $B(t) \equiv 0$ on any sub-interval $[t_i, t_{i+1}[$ (such that $t_i < t_{i+1}$) contained in $[t_r, t_s[$.

b) $s \geq r + k + 1$

c) I is maximal for this property.

Let us now set

$$\widetilde{B}(t) = \lambda_{r-k} B_{r-k}(t) + \dots + \lambda_{s-1} B_{s-1}(t).$$

The hypothesis c) implies that $B(t) \equiv 0$ on $[t_{r-1}, t_r[$. We have then $\lambda_{r-k} = \dots = \lambda_{r-1} = 0$ by 3.1.3; in the same way, $B(t) \equiv 0$ on $[t_s, t_{s+1}[$, which implies $\lambda_{s-k} = \dots = \lambda_s = 0$, whence

$$\widetilde{B}(t) = \lambda_r B_r(t) + \dots + \lambda_{s-k-1} B_{s-k-1}(t).$$

Moreover we have $t_r < u_r < \dots < u_{s-k-1} < t_s$ by hypothesis: the function $\widetilde{B}(t)$ has then at least $s - k - r$ distinct zeros inside the interval $[t_r, t_s[$ (as $\widetilde{B}(u_i) = B(u_i) = 0$ by hypothesis for $r \leq i \leq s-k-1$), and is not identically null on any sub-interval.

a) Assume that the zeros are simple (for the function $\widetilde{B}$).

Let $V(\widetilde{B})$ be the variation of the function $\widetilde{B}$ (see 1.5.5); we should then have $V(\widetilde{B}) \geq s - r - k$, which is in contradiction with the inequality $V(\widetilde{B}) \; \leq \; V(\lambda_r, \dots, \lambda_{s-k-1})$ (see 1.5.6), since we clearly have $V(\lambda_r, \dots, \lambda_{s-k-1}) \leq s - r - k - 1$.

b) General case

The function $\widetilde{B}(t)$ has at least $s-k-r$ distinct zeros inside the interval $[t_r, t_s[$, so it is immediately seen that for small enough $\epsilon > 0$, one of the two functions $\widetilde{B}(t)+\epsilon\sum_{j=r}^{s-k-1} B_j(t)$ or $\widetilde{B}(t)-\epsilon\sum_{j=r}^{s-k-1} B_j(t)$ has at least $s-r-k$ simple zeros in $[t_r, t_s[$. It is then enough to apply a) to this function.

■

3.1.4 Remark The same proof implies that under the hypotheses 3.1.1, all the principal minors of the matrix N (i.e., all the determinants of the matrix of the form $\left(B_j(u_i)\right)$ $(i = i_1, \ldots, i_r, \ j = i_1, \ldots, i_r))$ are non zero.

3.2 OTHER PROPERTIES OF THE MATRIX N

3.2.1 The matrix N is "sparse". More precisely, it is a "band" matrix, which has at most $k+1$ non zero elements in each row: the $(i+1)^{th}$ row of the matrix N is the row $B_j(u_i)$ $(0 \leq j \leq n-1)$; if $t_l \leq u_i \leq t_{l+1}$, only the B-splines $B_{l-k}, \ldots, B_l$ are possibly non zero at u_i.

3.2.2 Definition *A matrix M is totally positive if all its minors (i.e., the determinants of the square sub-matrices of M) are ≥ 0.*

A classical example of a totally positive matrix is the " Vandermonde matrix" $\left(u_i^{n_j}\right)$ for integers $n_1 < \cdots < n_p$ and real numbers u_i such that $0 < u_1 < \cdots < u_p$.

3.2.3 Proposition *The $n \times n$ matrix $N = \left(B_j(u_i)\right)$ defined above is totally positive.*

Proof Let $N_1 = \left(B_{r_j}(u_{r_i})\right)$ $(0 \leq i \leq s, \ 0 \leq j \leq s)$ be a sub-matrix of the matrix N, with determinant $|N_1|$. We have to prove that $|N_1| \geq 0$. Let us therefore assume that $|N_1| \neq 0$: this is equivalent to $B_{r_i}(u_{r_i}) \neq 0$ $(0 \leq i \leq s)$ (see Remark 3.1.4).

a) Assume that the knots t_i are simple

We have then $t_{r_i} < u_{r_i} < t_{r_i+k+1}$ $(0 \leq i \leq s)$, and $t_{r_i} < t_{r_i+1}$.

The function $|N_1|$ is a continuous function of the u_{r_i}'s, so one may make a continuous deformation of u_{r_i} into u'_{r_i} such that $t_{r_i} < u'_{r_i} < t_{r_i+1}$

in such a way that the determinant is always $\neq 0$ during the deformation. The matrix $N_1' = \left(B_{r_j}(u_{r_i}')\right)$ is then equal to

$$
\begin{pmatrix}
B_{r_1}(u_{r_0}') & 0 & \cdots & \cdots & 0 \\
B_{r_0}(u_{r_1}') & B_{r_1}(u_{r_1}') & 0 & \cdots & 0 \\
\vdots & \vdots & \ddots & & 0 \\
\vdots & \vdots & & & B_{r_s}(u_{r_s}')
\end{pmatrix}
$$

and $|N_1'| = \prod_{i=0}^{s} B_{r_i}(u_{r_i}')$ is strictly positive; thus one has also $|N_1| > 0$.

b) General case

Definition 1.3.2 implies that for fixed u, each function $B_{r_i}(u)$ is a continuous function of $t_0, \ldots, t_{n+k}$, with, as ever, the condition $t_i < t_{i+k+1}$ $(0 \leq i \leq n-1)$.

By the same argument as in a), we can continuously deform the knots t_i into knots $\tilde{t}_i$ close to t_i, such that $\tilde{t}_i < \tilde{t}_{i+1}$ $(0 \leq i \leq n-1)$ in such a way that the determinant of the matrix N_1 stays $\neq 0$ during the deformation. We may then apply the argument of a) to the matrix $\tilde{N}_1 = \left(\tilde{B}_{r_j}(u_i)\right)$, where the $\tilde{B}_{r_j}$ are the B-splines corresponding to the knot sequence $(\tilde{t}_i)$.

∎

3.2.4 Corollary   *Assume that the matrix $N = \left(B_j(u_i)\right)$ satisfies the conditions 3.1.1 (i.e., $|N| \neq 0$). One may then write $N = LU$, where L is an inferior triangular matrix, and U a superior triangular one.*

Proof   This results from the fact that the principal minors of N are all non zero: see [Ci], p. 83-84.

3.2.5 Remarks

1) The existence of the decompsition $N = LU$ is equivalent to the fact that the resolution of the linear system $NX = Y$ can be made without *pivoting*, which makes the resolution much easier, (using moreover the fact that N is a band matrix, see [Ci] or [DB]).

2) When the decomposition $N = LU$ is computed, the resolution of any linear system of the form $NX = Y$ is very fast: one has to solve successively the two following triangular systems

$$
\begin{cases}
LZ = Y \\
UX = Z
\end{cases}
$$

3) If the points P_i $(0 \leq i \leq n-1)$ are given, and we set $[a, b] = [0, 1]$, and if we want to determine a spline curve $B(t)$ of degree k, and points $\theta_i \in [0, 1]$ such that $B(\theta_i) = P_i$ $(0 \leq i \leq n-1)$, the simplest method consists of setting

$$L_i = |P_0 P_1| + \cdots + |P_{i-1} P_i| \qquad (1 \leq i \leq n-1)$$

and taking $\theta_i = L_i / L_n$ (so we take for θ_i the curvilinear abscissa of the piecewise linear spline passing through the points P_i).

One then deduces the value of the knots t_i, using 3.1.2, 3).

3.3 MATRIX REPRESENTATION

Let us look at a spline curve $B(t) = \sum_{i=0}^{n-1} P_i B_{i,k}(t)$ of degree k; by restricting $B(t)$ to the interval $[t_i, t_{i+1}[$ $(t_i < t_{i+1})$, we can identify it with a polynomial in t of degree $\leq k$; we will always assume in this section that the above interval is in fact $[0, 1[$; this can be obtained with a linear change of variable.

We may then write (on $[0, 1[$)

$$B(t) = \sum_{j=i-k}^{i} B_{j,k} P_j = (B_{i-k}(t) \quad \cdots \quad B_i(t)) \begin{pmatrix} P_{i-k} \\ \cdot \\ \cdot \\ P_i \end{pmatrix},$$

and each $B_{i,k}(t)$ is a polynomial in t, which may be written as its Taylor series at 0 (see the Differentiation algorithm 1.5.3); we may therefore write

$$(B_{i-k}(t) \quad \cdots \quad B_i(t)) = (1 \quad t \quad \cdots \quad t^k) M,$$

M being a $(k+1) \times (k+1)$ matrix; $B(t)$ is thus represented (on $[t_i, t_{i+1}[$) in the following matrix form

$$B(t) = \begin{pmatrix} 1 & t \ldots t^k \end{pmatrix} M \begin{pmatrix} P_{i-k} \\ \cdot \\ \cdot \\ P_i \end{pmatrix} \qquad (0 \leq t < 1).$$

We will give some examples of matrix representation, in the case (mostly used) where $k = 3$.

a) Uniform B-splines with integer knots

These B-splines are defined in Section 2.3, f) of Chapter 2.

Let us begin by computing the expression of $B_3(t)$ on each interval $[t_i, t_{i+1}[$ $(0 \leq i \leq 3)$, $B_3(t)$ being the B-spline of degree 3 with knots at 0,1,2,3,4. To do that, it is enough to apply the Definition 1.3.2

$$B_3(t) = B_{0,3}(t) = \frac{1}{3}\left(tB_{0,2}(t) + (4-t)B_{1,2}(t)\right).$$

$$\begin{cases} B_{0,2}(t) = \frac{1}{2}\left(tB_{0,1}(t) + (3-t)B_{1,1}(t)\right) \\ B_{1,2}(t) = \frac{1}{2}\left((t-1)B_{1,1}(t) + (4-t)B_{2,1}(t)\right). \end{cases}$$

$$\begin{cases} B_{0,1}(t) = tB_{0,0}(t) + (2-t)B_{1,0}(t) \\ B_{1,1}(t) = (t-1)B_{1,0}(t) + (3-t)B_{2,0}(t) \\ B_{2,1}(t) = (t-2)B_{2,0}(t) + (4-t)B_{3,0}(t). \end{cases}$$

$$B_{i,0}(t) = \begin{cases} 1 & \text{if } t_i \leq t < t_{i+1} \\ 0 & \text{otherwise,} \end{cases}$$

which gives

$$B_3(t) = \begin{cases} \dfrac{t^3}{6} & \text{for } 0 \leq t < 1 \\[2mm] \dfrac{1}{6}(-3t^3 + 12t^2 - 12t + 4) & \text{for } 1 \leq t < 2 \\[2mm] \dfrac{1}{6}(3t^3 - 24t^2 + 60t - 44) & \text{for } 2 \leq t < 3 \\[2mm] \dfrac{1}{6}(4-t)^3 = \dfrac{1}{6}(-t^3 + 12t^2 - 48t + 64) & \text{for } 3 \leq t < 4. \end{cases}$$

Let us now look at a uniform spline curve $\sum_i B_3(t-i)P_i$ defined on $\mathbf{R}$. On the interval $[0,1[$, this sum is reduced to

$$B_3(t+3)P_{-3} + B_3(t+2)P_{-2} + B_3(t+1)P_{-1} + B_3(t)P_0$$

$$= \frac{1}{6}\left[(1-t)^3 P_{-3} + \left(3(t+2)^3 - 24(t+2)^2 + 60(t+2) - 44\right)P_2\right.$$

$$\left. + \left(-3(t+1)^3 + 12(t+1)^2 - 12(t+1) + 4\right)P_{-1} + t^3 P_0\right]$$

$$= \frac{1}{6}\Big[(-t^3 + 3t^2 - 3t + 1)P_{-3} + (3t^3 - 6t^2 + 4)P_{-2}$$

$$+(-5t^3 + 3t^2 + 3t + 1)P_{-1} + t^3 P_0\Big],$$

which gives the following matrix representation

$$B(t) = \frac{1}{6}\begin{pmatrix} 1 & t & t^2 & t^3 \end{pmatrix}\begin{pmatrix} 1 & 4 & 1 & 0 \\ -3 & 0 & 3 & 0 \\ 3 & -6 & 3 & 0 \\ -1 & 3 & -3 & 1 \end{pmatrix}\begin{pmatrix} P_{-3} \\ P_{-2} \\ P_{-1} \\ P_0 \end{pmatrix}$$

for $0 \le t < 1$.

The portion of the curve $B(t)$ corresponding to $i \le t < i+1$ is written with the same matrix

$$B(\theta) = \frac{1}{6}\begin{pmatrix} 1 & \theta & \theta^2 & \theta^3 \end{pmatrix}\begin{pmatrix} 1 & 4 & 1 & 0 \\ -3 & 0 & 3 & 0 \\ 3 & -6 & 3 & 0 \\ -1 & 3 & -3 & 1 \end{pmatrix}\begin{pmatrix} P_{i-3} \\ P_{i-2} \\ P_{i-1} \\ P_i \end{pmatrix} \quad (0 \le \theta < 1).$$

b) Bézier curves (see Chapter 2, Section 2.1)

Let P_i $(0 \le i \le 3)$ be points in $\mathbf{R}^s$; we have

$$f(t) = P_0 B_0^3(t) + P_1 B_1^3(t) + P_2 B_2^3(t) + P_3 B_3^3(t),$$

where the $B_i^3(t)$'s are the Bernstein polynomials of degree 3 (see 2.1.2), whence

$$f(t) = \frac{1}{6}\begin{pmatrix} 1 & t & t^2 & t^3 \end{pmatrix}\begin{pmatrix} 1 & 0 & 0 & 0 \\ -3 & 3 & 0 & 0 \\ 3 & -6 & 3 & 0 \\ -1 & 3 & -3 & 1 \end{pmatrix}\begin{pmatrix} P_0 \\ P_1 \\ P_2 \\ P_3 \end{pmatrix}$$

$(0 \le t \le 1)$.

c) "Hermite's curves"

For the sake of completeness, we will give the case of Hermite curves, which are not B-splines curves in the above sense.

With four points P_0, P_1, P_2, P_3 given, we look for a parametric curve $\gamma_1(t)$ of degree 3 $(0 \leq t \leq 1)$ such that (see Figure 3.3.1)

$$\begin{cases} \gamma_1(0) = P_1 \\ \gamma_1(1) = P_2 \\ \gamma_1'(0) = P_0 - P_2 \\ \gamma_1'(1) = P_1 - P_3. \end{cases}$$

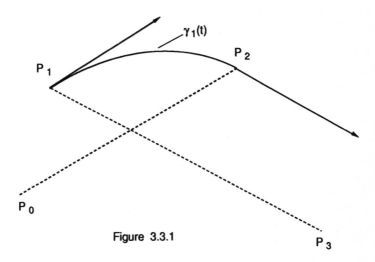

Figure 3.3.1

If a sequence P_i $(0 \leq i \leq n - 1)$ is given, such a curve allows the interpolation of these points by a curve of class $\mathcal{C}^1$, made with pieces γ_i verifying

$$\begin{cases} \gamma_i(0) = P_i \\ \gamma_i(1) = P_{i+1} \\ \gamma_i'(0) = P_{i-1} - P_{i+1} \\ \gamma_i'(1) = P_i - P_{i+2}. \end{cases}$$

Setting $\gamma_1(t) = Q_0 + Q_1 t + Q_2 t^2 + Q_3 t^3$ we have the following relations

$$\begin{cases} Q_0 = P_1 \\ Q_0 + Q_1 + Q_2 + Q_3 = P_2 \\ Q_1 = P_0 - P_2 \\ Q_1 + 2Q_2 + 3Q_3 = P_1 - P_3, \end{cases} \quad \text{or} \quad \begin{cases} Q_0 = P_1 \\ Q_1 = P_0 - P_2 \\ Q_2 = -2P_0 - 4P_1 + 5P_2 + P_3 \\ Q_3 = P_0 + 3P_1 - 3P_2 - P_3, \end{cases}$$

which gives the following matrix representation

$$\gamma_1(t) = \left(1, t, t^2, t^3\right) \begin{pmatrix} 0 & 1 & 0 & 0 \\ 1 & 0 & -1 & 0 \\ -2 & -4 & 5 & 1 \\ 1 & 3 & -3 & -1 \end{pmatrix} \begin{pmatrix} P_0 \\ P_1 \\ P_2 \\ P_3 \end{pmatrix}.$$

3.4 JUNCTION BETWEEN TWO CURVES

a) Case of spline curves

Let us consider two spline curves of degree k in $\mathbf{R}^s$:

$$S(t) = \sum_{i=0}^{n-1} B_{i,k}(t) P_i$$

defined on $[a, b]$ with knots $t = (t_0, \ldots, t_{n+k})$, and

$$S'(t) = \sum_{j=0}^{n'-1} B_{j,k} P_j'$$

defined on $[a', b']$ with knots $t' = (t'_0, \ldots, t'_{n'+k})$.

If we want to join S and S', it is most convenient to assume $t'_0 = t_n, \ldots, t'_k = t_{n+k}$ (recall that we have $t_i \geq b$ for $n \leq i \leq n + k$ and $t'_j \leq a'$ for $0 \leq j \leq k$). If one sets

$$\begin{cases} Q_i = P_i & (0 \leq i \leq n - 1) \\ Q_{n+j} = P_j' & (0 \leq j \leq n' - 1), \end{cases}$$

the curve

$$\Gamma(t) = \sum_{i=0}^{n+n'-1} B_{i,k}(t) Q_i$$

is a spline curve which coincides with $S(t)$ for $t \in [a, b[$, and with $S'(t)$ for $t \in [a', b']$ (see Figure 3.4.1).

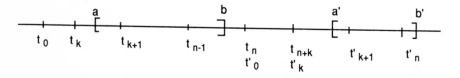

Figure 3.4.1

3.4.1 Remarks

1) The junction is clearly simpler to make in the case of spline curves with uniform knots.

2) If we require $b = a'$, we have to set $t_n = \cdots = t_{n+k} = b$ and $t'_0 = \cdots = t'_k = a' = b$; this will for instance be the case for Bézier curves, which we will study below; if moreover we want a continuous junction between the two curves, we then have to assume that $P_{n-1} = P'_0$. The junction will then in general be only C^0 at P_{n-1}, as the spline curve $\Gamma(t)$ will have at this point a knot of order $k + 1$.

We will study the case of a differentiable junction in the case of Bézier curves, leaving to the reader the analogous case of spline curves (for which the question arises only when one wants $a' = b$ and $S(b) = S'(a')$).

b) Case of Bézier curves

Let $B(t) = \sum_{i=0}^{k} B_i^k(t) P_i$ $(0 \le t \le 1)$ be a Bézier curve of degree k, $B_i^k(t)$ being the ith Bernstein polynomial (see 2.1.1). Set

$$\Delta P_i = P_{i+1} - P_i$$

$$\Delta^r P_i = \Delta(\Delta^{r-1} P_i) = \sum_{j=0}^{r} (-1)^{r-j} \binom{r}{j} P_{i+j}.$$

If D denotes the derivation operator, we then have (see 2.3.5)

$$\begin{cases} DB(t) = k \sum_{i=0}^{k-1} \Delta(P_i) B_i^k(t) \\ \\ D^r B(t) = \dfrac{k!}{(k-r)!} \sum_{i=0}^{k-r} \Delta^k(P_i) B_i^{k-r}(t). \end{cases}$$

If now $B'(t) = \sum_{i=0}^{k} B_i^k(t-1) P'_i$ is another Bézier curve, of the same degree k, defined for $1 \le t \le 2$, we can then join the curve B' to the curve B at the point $t = 1$ under the following conditions:

C^0 continuity : $\quad P_k = P'_0$.

C^1 continuity: the same condition, plus $\Delta P_{k-1} = \Delta P'_0$, or $P_k - P_{k-1} = P'_1 - P'_0$.

C^2 continuity: the same conditions, plus $\Delta^2 P_{k-2} = \Delta^2 P'_0$, or

$$P_k - 2P_{k-1} + P_{k-2} = P'_2 - 2P'_1 + P'_0,$$

or else

$$P_2' - P_{k-2} = 2(P_1' - P_{k-1}) \qquad \text{since } P_k = P_0').$$

C^p continuity: $\Delta^i P_{k-i} = \Delta^i P_0'$ $(0 \le i \le p)$.

Examples

a) C^2 continuity between two Bézier curves

$$\begin{cases} |P_{k-1}P_k| = |P_kP_1'| \\ |P_{k-2}P_2'| = 2|P_{k-1}P_1'| \end{cases}$$

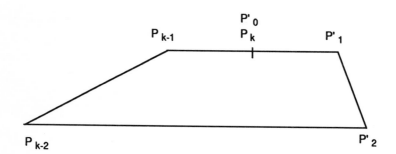

Figure 3.4.2

b) A closed Bézier curve of degree 5 and of class C^2: $P_5 = P_0$, $|P_1P_0| = |P_0P_4|$, $|P_2P_3| = 2|P_1P_4|$.

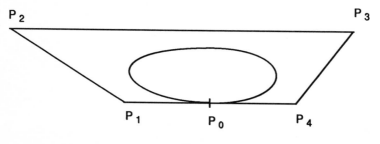

Figure 3.4.3

3.4.2 Remarks

1) The condition of C^2 continuity implies in particular that the points $P_{k-2}, P_{k-1}, P_k = P_0', P_1'$ and P_2' are coplanar.

2) If $P_k = P_0'$ and if P_{k-1}, P_k and P_1' are on a straight line without the condition $|P_{k-1}P_k| = |P_kP_1'|$, there is a discontinuity in the modulus of the tangent (at P_k), but not in its direction. In other words, there will be a junction with C^1 continuity after a (for instance linear) change of parameters on one of the two curves (which does not change the image of the curve in $\mathbf{R}^s$). The discontinuity of the tangent is thus invisible on the image of the curve (since only the "speed" has a discontinuity).

One says then that there is *visual continuity* of the tangent, or that the junction at P_k is *visually* C^1. One sometimes also says that there is G_1 *continuity* or *geometric continuity* at the point P_k (see e) below).

There are similar notions for more differentiable junctions, for instance the continuity of the curvature (see d) below).

c) Degree elevation in the case of Bézier curves

If the two Bézier curves $B^k(t)$ and $B^{k'}(t)$ are not of the same degree, (with for instance $k < k'$), one can turn to the case $k = k'$ by "elevating the degree" of $B^k(t)$ in the following way. Assuming, by induction, that $k' = k + 1$, one may consider the curve $B^k(t) = \sum_{i=0}^{k} P_i B_i^k(t)$ as a special curve of degree $\leq k + 1$

$$B^k(t) = \sum_{i=0}^{k+1} Q_i B_i^{k+1}(t).$$

3.4.3 Proposition   *With the above notation, we have*

$$\begin{cases} Q_0 = P_0 \\ Q_{k+1} = P_k \\ Q_i = \dfrac{iP_{i-1} + (k - i + 1)P_i}{k + 1} \quad (1 \leq i \leq k). \end{cases}$$

Proof   It suffices to remark that we can write

$$B_i^k(t) = (1 - t)B_i^k(t) + tB_i^k(t) = \left(\frac{k + 1 - i}{k + 1}\right)B_i^{k+1}(t) + \left(\frac{i + 1}{k + 1}\right)B_{i+1}^{k+1}(t)$$

(see 2.1.2), and to replace in the expression

$$B^k(t) = \sum_{i=0}^{k} B_i^k(t) P_i.$$

■

Example

Degree elevation in the case of a Bézier curve of degree 3 (see Figure 3.4.4; note that the operation "degree elevation" moves the control polygon closer to the curve). We have

$$Q_1 = \frac{1}{4}(P_0 + 3P_1), \quad Q_2 = \frac{1}{2}(P_1 + P_2), \quad Q_3 = \frac{1}{4}(3P_2 + P_3).$$

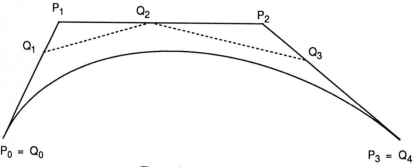

Figure 3.4.4

3.4.4 Remark   In the case of a spline curve $S_k(t)$ of degree k, the operation "degree elevation" does not have a meaning as clear as above. If for instance the knots $(t_i)_{k+1 \leq i \leq n-1}$ are simple, the curve $S_k(t)$ is of class C^{k-1}: one cannot in general look to it as a special spline curve $S_{k+1}(t)$ (with the same knots), as any spline curve of degree $k + 1$ and defined with simple knots is of class C^k. If we want to join the curve $S_k(t)$ with a spline curve of degree $k + 1$, one solution consists in considering the portion of $S_k(t)$ for $t_n \leq t < t_{n+1}$ (which is then of degree $\leq k$) as a Bézier curve of degree k, and to apply the above method.

d) Curvature of a Bézier curve

In the same spirit as above, let us point out how the curvature of a Bézier curve is related to its control polygon (for more details, see [F-P]).

Recall that if $B(t)$ is a curve in $\mathbf{R}^s$ of class $\mathcal{C}^2$, its curvature at the point $B(t_0)$ is given by the formula

$$K(t_0) = \frac{\left|B'(t_0) \wedge B''(t_0)\right|}{\left|B'(t_0)\right|^3},$$

$B' \wedge B''$ denoting the vector product of the vectors B' and B''.

In the case of a Bézier curve of degree k: $B(t) = \sum_{i=0}^{k} B_i^k(t)P_i$, we have therefore

$$\begin{cases} K(0) = \dfrac{k-1}{k} \dfrac{\left|(P_1 - P_0) \wedge (P_2 - P_1)\right|}{\left|P_1 - P_0\right|^3} \\[2ex] K(1) = \dfrac{k-1}{k} \dfrac{\left|(P_{k-1} - P_k) \wedge (P_{k-2} - P_{k-1})\right|}{\left|P_{k-1} - P_k\right|^3} \end{cases}$$

since for instance (see 2.3.5)

$$\begin{cases} B'(0) = k(P_1 - P_0) \\ B''(0) = k(k-1)\left[P_0 - 2P_1 + P_2\right] \\ \qquad\;\; = k(k-1)\left[(P_2 - P_1) - (P_1 - P_0)\right]. \end{cases}$$

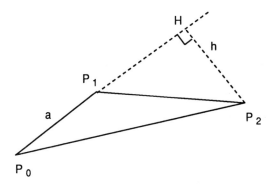

Figure 3.4.5

Figure 3.4.5 shows a geometric interpretation of the curvature: $K(0) = \frac{2(k-1)}{k} \times \left(\frac{h}{a^2}\right)$, with $a = |P_1 - P_0|$, $h = |P_2 - H|$.

e) Geometric continuity

3.4.5 Definition   *Let r be an integer ≥ 0, $S_1(t)$ $(0 \leq t \leq 1)$ and $S_2(t)$ $(1 \leq t \leq 2)$ two parametric curves of class C^r in $\mathbf{R}^s$ such that $S_1(1) = S_2(1)$, and $S_1'(1) \neq 0$ or $S_2'(1) \neq 0$; if i is an integer such that $1 \leq i \leq r$, one says that there is a G_i junction (or G_i continuity) at the point $S_1(1) = S_2(1)$ if there exists a change of parameter of class C^i $t = \phi(u)$ $(0 \leq u \leq 1$, $\phi(1) = 1)$, $\phi'(1) > 0$ such that the curve $S_1(\phi(u))$ has a C^i junction with $S_2(t)$ at $u = t = 1$.*

3.4.6   Let us point out the conditions for a G_i junction for $0 \leq i \leq 2$. Notation is the same as in 3.4.5.

  a) The G_0 junction condition means that $S_1(1) = S_2(1)$: it is thus equivalent to the C^0 junction condition.

  b) There is G_1 junction if and only if there is G_0 junction, and if there exists a constant $\beta_1 > 0$ such that $\beta_1 S_1'(1) = S_2'(1)$ (one has $\beta_1 = \phi'(1)$ with the above notation).

  The G_1 continuity means then that the tangents to the two curves (at the point corresponding to $t = 1$) have the same direction and the same orientation.

  The C^1 continuity case corresponds to the case $\beta_1 = 1$.

  c) There is G_2 continuity if and only if there is G_1 continuity, and if there exists a constant β_2 such that

$$(3.4.6, 1) \qquad\qquad S_2''(1) = \beta_2 S_1'(1) + \beta_1^2 S_1''(1).$$

Let us prove this last formula; we compute the derivatives of $T(u) = S_1(\phi(u))$ at $u = 1$

$$\begin{cases} T'(1) = S_1'(1)\phi'(1) = \beta_1 S_1'(1) \\ T''(1) = \beta_1^2 S_1''(1) + \phi''(1)S_1'(1). \end{cases}$$

Setting $\beta_2 = \phi''(1)$, we find the above formula (3.4.6, 1); the C^2 case corresponds to $\beta_1 = 1$, $\beta_2 = 0$.

3.4.7 Remarks   1) The G_2 continuity has the same geometric meaning as the C^2 one: the curves $S_1(t)$ and $S_2(t)$ have the same image in $\mathbf{R}^s$. They are only covered with different "speeds". The advantage of the G_2 condition is that it introduces two additional parameters, and so more flexibility.

2) Traditionally, β_1 is called parameter of *bias* and β_2 parameter of *tension*.

3.4.8   As an example, we will describe the uniform "β-splines" of degree three, following Barsky ([Ba]).

Let a uniform knot sequence be given (for instance $t_i = i$), along with two constants β_1 and β_2, with $\beta_1 > 0$. We want to generalize the B-splines of degree 3, replacing them by piecewise polynomial functions $S_i(t)$, with G_2 junctions at the knots t_i, the parameters of bias and tension being respectively equal to β_1 and β_2 at each t_i. Since the knot sequence is uniform, we have only to define the function $S_0(t)$, whose support will be $[0,4]$. It is equivalent (and often more convenient), to determine the matrix

$$M = \left(a_{i,j} \right)_{0 \leq i \leq 2, \, -3 \leq j \leq 0}$$

such that if we set $S(t) = \sum_j P_j S_j(t)$, we have:

$$S(t) = \left(1 \; t \; t^2 \; t^3 \right) M \begin{pmatrix} P_{-3} \\ P_{-2} \\ P_{-1} \\ P_0 \end{pmatrix}. \quad \text{So}$$

$$S(t) = \sum_j P_j S_j(t) = \sum_{j=-3}^{0} P_j S_0(t - j),$$

and we set

$$S_0(t - j) = S_j(t) = a_{o,j} + a_{1,j}t + a_{2,j}t^2 + a_{3,j}t^3.$$

Recall from 3.3, a), that in the B-spline case ($\beta_1 = 1$, $\beta_2 = 0$), we had:

$$M = \frac{1}{6} \begin{pmatrix} 1 & 4 & 1 & 0 \\ -3 & 0 & 3 & 0 \\ 3 & -6 & 3 & 0 \\ -1 & 3 & -3 & 1 \end{pmatrix}.$$

We then need a system of 16 linear equations to determine the $a_{i,j}$'s. Let us describe these equations.

a) Continuity of $S_0(t)$ at the points 0,1,2,3,4.
We find the five following equations

$$\begin{cases} a_{0,0} = 0 \quad \text{(continuity at 0)} \\ a_{0,0} + a_{1,0} + a_{2,0} + a_{3,0} = a_{0,-1} \quad \text{(continuity at 1)} \\ a_{0,-1} + a_{1,-1} + a_{2,-1} + a_{3,-1} = a_{0,-2} \\ a_{0,-2} + a_{1,-2} + a_{2,-2} + a_{3,-2} = a_{0,-3} \\ a_{0,-3} + a_{1,-3} + a_{2,-3} + a_{3,-3} = 0. \end{cases}$$

b) The G_1 and G_2 continuity conditions at the points 0,1,2,3,4 give respectively the two following systems

$$\begin{cases} a_{1,0} = 0 \\ \beta_1(a_{1,0} + 2a_{2,0} + 3a_{3,0}) = a_{1,-1} \\ \beta_1(a_{1,-1} + 2a_{2,-1} + 3a_{3,-1}) = a_{1,-2} \\ \beta_1(a_{1,-2} + 2a_{2,-2} + 3a_{3,-2}) = a_{1,-3} \\ a_{1,-3} + 2a_{1,-3} + 3a_{3,-3} = 0 \end{cases} \qquad \begin{cases} a_{2,0} = 0 \\ \beta_1^2 a_{2,0} + \beta_2 a_{1,0} = a_{2,-1} \\ \beta_1^2 a_{2,-1} + \beta_2 a_{1,-1} = a_{2,-2} \\ \beta_1^2 a_{2,-2} + \beta_2 a_{1,-2} = a_{2,-3} \\ \beta_1^2 a_{2,-3} + \beta_2 a_{1,-3} = 0. \end{cases}$$

This yields a total of 15 equations in 16 unknowns; we add the equation $\sum_{j=-3}^{0} S_j(0) = 1$, i.e., $a_{0,-3} + a_{0,-2} + a_{0,-1} + a_{0,0} = 1$ to get a system whose determinant is

$$\Delta = \beta_2 + 2\beta_1^3 + 4\beta_1^2 + 4\beta_1 + 2$$

which we assume is non zero (which is the case if for instance $\beta_1 > 0$ and $\beta_2 \geq 0$).

We find then a unique solution for the $a_{i,j}$'s, which gives a set of base functions generalizing the uniform B-splines of degree three, which we call the β-splines (cf [Ba]). The matrix $(a_{i,j})$ thus equals

$$M = \frac{1}{\Delta} \begin{pmatrix} 2\beta_1^2 & \beta_2 + 4\beta_1^2 + 4\beta_1 & 2 & 0 \\ -6\beta_1^3 & 6\beta_1^3 - 6\beta_1 & 6\beta_1 & 0 \\ 6\beta_1^3 & -(3\beta_2 + 6\beta_1^3 + 6\beta_1^2) & 3\beta_2 + 6\beta_1^2 & 0 \\ -2\beta_1^3 & 2\beta_2 + 2\beta_1^3 + 2\beta_1^2 + 2\beta_1 & -(2\beta_2 + 2\beta_1^2 + 2\beta_1 + 2) & 2 \end{pmatrix}.$$

It is easy to verify that $\sum S_j(t) = 1$, which implies the convex hull property for the curves constructed with these base functions.

Let us look at the influence of the tension parameter β_2; when β_2 tends to $+\infty$ and $\beta_1 = 1$, the base functions become

$$\begin{cases} S_{-3}(t) = 0 \\ S_{-2}(t) = 1 - (3t^2 - 2t^3) \\ S_{-1}(t) = 3t^2 - 2t^3 \\ S_0(t) = 0 \end{cases} \quad (0 \le t \le 1).$$

The curve $S(t) = \sum S_j(t)P_j$ has then the expression $(1-\alpha)P_{-2} + \alpha P_{-1}$ for $0 \le t \le 1$, with $\alpha = 3t^2 - 2t^3$. Its image is the segment $P_{-2}P_{-1}$. We see that when $\beta_2 \longrightarrow +\infty$, the curve $S(t)$ tends to be merged with the control polygon (always staying G_2), which explains the name tension given to the parameter β_2.

The bias parameter is not so clear to analyze; it controls the shape of the curve at the junction points between two polynomial pieces. If β_1 is big, the curve is closer to the tangent at the "right side" of the knot than at the "left one".

3.5 RATIONAL CURVES

It is impossible to represent an arc of a conic $C \subset \mathbf{R}^2$ which is not a parabola under the parametric form

$$(3.5.1, 1) \qquad \begin{cases} X = g(t) \\ Y = h(t) \end{cases}$$

where g and h are polynomials in t; this comes for instance from the fact that such a curve has only one "infinite point" (obtained for "$t = \infty$"), whereas a hyperbola has two, and an ellipse zero (on the field $\mathbf{R}$), or two (on the field $\mathbf{C}$).

For instance, if the circle $X^2 + Y^2 = 1$ could contain an arc satisfying (3.5.1, 1), we would have, on an interval of $\mathbf{R}$, $g^2(t) + h^2(t) = 1$, which is clearly impossible; the same is so for the hyperbola $XY - 1 = 0$.

However the circle $X^2 + Y^2 = 1$ can be parametrized with rational fractions

$$(3.5.1, 2) \qquad \begin{cases} X = \dfrac{1 - t^2}{1 + t^2} \\ Y = \dfrac{2t}{1 + t^2}. \end{cases}$$

This motivates the introduction of rational fractions in the definition of spline functions. We will begin by recalling those curves in $\mathbf{R}^2$ that have a parametrization by rational fractions.

a) Rational curves

3.5.1 Definition   *An algebraic curve $\Gamma \subset \mathbf{R}^2$ (defined by an equation $P(X,Y) = 0$, $P \in \mathbf{R}[X,Y]$) is called rational if it admits a rational parametrization of the form:*

$$(3.5.1, 3) \qquad\qquad \begin{cases} X = g(t) \\ Y = h(t), \end{cases}$$

where g and h are rational fractions.

3.5.2 Remarks

1) If we want all the points of Γ under the form $(3.5.1, 3)$, it is sometimes necessary to admit the value "∞" for t (which gives for instance the point $(-1,0)$ of the circle $X^2 + Y^2 = 1$ parametrized by $(3.5.1, 2)$). In another way, some values of t (for which for instance the denominator is zero) may correspond to "points at infinity" of Γ.

We will not formalize these notions of elementary algebraic geometry (which motivate the introduction of projective spaces), that are very easy and intuitive in the cases considered here.

2) One can make similar considerations for (not necessarily plane) curves in $\mathbf{R}^s$: we have considered only the $s = 2$ case due to the difficulty of the definition of a general algebraic curve.

3.5.3 Definition   *Let $\Gamma \subset \mathbf{R}^2$ be an algebraic curve of degree d (i.e., defined by an equation $P(X,Y) = 0$, with $P \in \mathbf{R}[X,Y]$ of degree d).*

α) One says that the origin $0 \in \Gamma$ is a point of order $d - 1$, if the equation $P(X,Y) = 0$ of Γ is of the form

$$(3.5.3, 1) \qquad\qquad P_{d-1}(X,Y) + P_d(X,Y) = 0$$

where P_{d-1} (resp. P_d) is homogeneous of degree $d - 1$ (resp. d); if $d > 1$, one says moreover that 0 is singular.

β) One says that a point $Q \in \Gamma$ is of order $d-1$, if, after a linear change of coordinates which sends Q to 0, α) above is satisfied.

3.5.4 Proposition

a) *Any non empty smooth conic (i.e., not equal to the union of two distinct (or otherwise) straight lines) $C \subset \mathbf{R}^2$ is rational.*

b) *If $\Gamma \subset \mathbf{R}^2$ is a curve of degree d having a point Q of order $d-1$ and containing no straight line passing by Q, it is rational.*

Proof We see that a) is a particular case of b), taking for Q any point of C, as then $d = 2$. For the proof of b), set $Y = tX$; equation (3.5.3,1) becomes then

$$X^{d-1}\left[P_{d-1}(1,t) + XP_d(1,t)\right] = 0.$$

We find in this way the solutions $(X = 0, Y = 0)$ which correspond to the origin "counted $d-1$ times", and

$$\begin{cases} X = -\dfrac{P_{d-1}(1,t)}{P_d(1,t)} \\[2mm] Y = tX = -\dfrac{tP_{d-1}(1,t)}{P_d(1,t)}, \end{cases}$$

which gives a rational parametrization of Γ.

∎

For completeness, we recall (without going into details), the following notions (see for instance [Fu]): one can define the *genus* of an algebraic irreducible curve Γ (i.e., defined by an equation $P(X,Y) = 0$ where P is an irreducible polynomial over the field $\mathbf{C}$ of complex numbers). We have in particular the following facts:

- if Γ is without singular point, ("real or complex, at finite or infinite distance"), and P of degree d, we have $g = \frac{(d-1)(d-2)}{2}$.

- if P is of degree d and Γ has k singular ordinary quadratic points (real or complex, at finite or infinite distance), $g = \frac{(d-1)(d-2)}{2} - k$.

3.5.5 Proposition   Let $\Gamma \subset \mathbf{R}^2$ be an irreducible algebraic curve; then Γ is rational if and only if it is of genus 0.

We see for instance that the only rational smooth curves are straight lines and the conics, and that the cubics are rational only if they have a singular point (which, for reasons of degree, could only be a cusp or an ordinary quadratic point).

b) Rational splines

We shall always keep the same notation; let $B_{i,k,t}$ $(0 \leq i \leq n-1)$ the B-splines of degree k on $[a,b]$, corresponding to a knot sequence $t = (t_0, \ldots, t_{n+k})$.

3.5.6 Definition   Let $P_0, \ldots, P_{n-1}$ be points in $\mathbf{R}^s$, $w_0, \ldots, w_{n-1}$ n real numbers not all zero. The rational B-spline curve associated to this data is the parametric curve Γ defined by

$$S(t) = \frac{\sum_{i=0}^{n-1} P_i w_i B_i(t)}{\sum_{i=0}^{n-1} w_i B_i(t)}.$$

The w_i's are the *weights* associated to the curve Γ; one restricts in general the range of t to an interval $[a,b]$ such that $t_k \leq a$, $t_n \geq b$, which we will assume in the sequel.

3.5.7 Properties of the rational spline curves

1) One may write $S(t) = \sum_{i=0}^{n-1} P_i \psi_i(t)$, where $\psi_i(t) = \dfrac{w_i B_i(t)}{\sum_{i=0}^{n-1} w_i B_i(t)}$ is a rational fraction; Γ is then a rational curve as defined above (see 3.5.1).

2) If $w_i = 1$ $(0 \leq i \leq n-1)$, we find the usual notion of spline curve, when the range of t is restricted to $[a,b]$ $(t_k \leq a, t_n \geq b)$, as then we have the relation $\sum_{i=0}^{n-1} B_{i,k}(t) = 1$ $\forall t \in [a,b]$ (see 1.3.4).

3) If $w_i > 0$ $(0 \leq i \leq n-1)$, the denominator of $S(t)$ is not zero in $[a,b]$; this is the most frequent case.

4) We have $\sum_{i=0}^{n-1} \psi_i(t) = 1$ $(t \in [a,b[)$, and then, as for polynomial splines in the case where the w_i's are > 0, the curve Γ is situated in the convex hull of the points P_i; as we have $\psi_i(t) = \dfrac{w_i B_i(t)}{\sum_{i=0}^{n-1} w_i B_i(t)}$, we have moreover the same properties for the support than for the polynomial splines (see 1.3.4).

5) Let us take for instance $s = 3$, i.e., $P_i \in \mathbf{R}^3$. Let $\tilde{\Gamma} \subset \mathbf{R}^4 = \mathbf{R}^3 \times \mathbf{R}$ be the curve defined by the parametrization

$$\tilde{S}(t) = \begin{cases} \sum_{i=0}^{n-1} w_i P_i B_i(t), \\ \sum_{i=0}^{n-1} w_i B_i(t). \end{cases}$$

The curve $\widetilde{\Gamma}$ is therefore the B-spline in $\mathbf{R}^4$ having for its control polygon the set of points

$$Q_i = (w_i P_i, w_i) \in \mathbf{R}^3 \times \mathbf{R} \quad (0 \leq i \leq n-1).$$

Let X_1, X_2, X_3, X_4 be coordinates on $\mathbf{R}^4$; Γ is then the image of $\widetilde{\Gamma}$ by the map

$$\pi: \quad \mathbf{R}^4 \setminus \{X_4 = 0\} \longrightarrow \mathbf{R}^3$$
$$(X_1, X_2, X_3, X_4) \longmapsto (X_1/X_4, X_2/X_4, X_3/X_4).$$

We may then apply to $\widetilde{\Gamma}$ all the algorithms of the preceding chapters, which will give algorithms for Γ using the projection π.

6) When the curve $\widetilde{\Gamma}$ passes through the origin of $\mathbf{R}^4$ (or, more generally, if the denominator and the numerator of one of the coordinates of $S(t)$ are simultaneously zero for the value t_0 of t), the corresponding point of Γ is indeterminate (one says sometimes that it is a *base point*). This question will be studied in the case of surfaces in the next chapter (Section 7).

This happens for instance for the rational Bézier curves (see below) when $w_0 = 0$, for $t = 0$ (see next chapter).

c) Special case of Bézier curves

Recall that the ith Berstein polynomial of degree k on $[0,1]$ is defined by $B_i^k(t) = \binom{k}{i} t^i (1-t)^{k-i}$ $(0 \leq i \leq k)$; in the same manner, $B_i^k \left(\frac{t-a}{b-a}\right)$ is the ith Bernstein polynomial on $[a, b]$ (see 2.1.2 and 2.1.5).

A rational Bézier curve is thus defined as above by

$$B(t) = \frac{\sum_{i=0}^{k} w_i P_i B_i^k(t)}{\sum_{i=0}^{k} w_i B_i^k(t)} \quad (0 \leq t \leq 1).$$

We may then, without modifying the curve, make a *homographic* transformation on the parameter: $t = \dfrac{a\theta + b}{c\theta + d}$ (with $ad - bc \neq 0$); ($t^i(1-t)^{k-i}$ is then transformed to $\dfrac{(a\theta + b)^i ((c-a)\theta + d - b)^{k-i}}{(c\theta + d)^k}$), which may sometimes simplify the expression for $B(t)$, and lets us fix the values of the parameter θ for three given points on the curve (the transformation $t \mapsto \dfrac{a\theta + b}{c\theta + d}$ depends on three parameters, as a, b, c, d are defined modulo a nonzero multiplicative constant).

For instance, if the "value" $\theta = \infty$ corresponds to a real point "at infinity", the degree of the denominator becomes $\leq k - 1$.

d) Some properties of rational representations of conics

This section follows [L]. Let us look at Bézier curves of degree two; we have

$$\begin{cases} B_0^2(t) = (1-t)^2 \\ B_1^2(t) = 2t(1-t) \\ B_2^2(t) = t^2. \end{cases}$$

A rational Bézier curve of degree two is then of the form

$$(3.5.7, 1) \qquad \gamma(t) = \frac{w_0 P_0 (1-t)^2 + 2w_1 P_1 t(1-t) + w_2 P_2 (t^2)}{w_0 (1-t)^2 + 2w_1 t(1-t) + w_2 t^2},$$

where w_0, w_1 and w_2 are real numbers not all zero (see Figure 3.5.1).

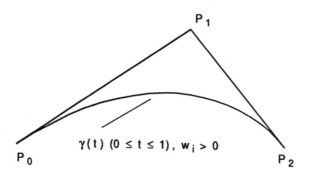

Figure 3.5.1

3.5.8 Properties of the rational Bézier curves of degree two

Assume $w_i > 0$ $(0 \leq i \leq 2)$. Then

1)
$$\gamma(0) = P_0, \qquad \gamma(1) = P_2,$$

$$\gamma'(0) = \frac{2w_1}{w_0}(P_1 - P_0), \qquad \gamma'(1) = \frac{2w_1}{w_0}(P_2 - P_1).$$

2) $\gamma(t)$ is an arc of conic C; more precisely, let

$$\gamma(t) = P_0 \psi_0(t) + P_1 \psi_1(t) + P_2 \psi_2(t)$$

and let us consider the coordinate sytem with origin P_1, and with unit vectors $\overrightarrow{u} = P_0 - P_1$, $\overrightarrow{v} = P_2 - P_1$. We may write

$$\gamma(t) = \psi_0(t)(P_0 - P_1) + \psi_2(t)(P_2 - P_1) + P_1$$

(since $\sum_{i=0}^2 \psi_i(t) = 1$), and then the coordinates of $\gamma(t)$ (in this system) are

$$\begin{cases} \alpha = \psi_0(t) \\ \beta = \psi_2(t). \end{cases}$$

Set $w(t) = w_0(1-t)^2 + 2w_1 t(1-t) + w_2 t^2$; we have

$$\alpha\beta = \frac{w_0 w_2}{(w(t))^2} B_0(t) B_2(t) = \frac{w_0 w_2}{(w(t))^2} \frac{(B_1^2(t))^2}{4} = \frac{w_0 w_2}{4 w_1^2} (\psi_1(t))^2,$$

which gives the intrinsic equation of C (always in the same system of coordinates)

$$(3.5.8, 1) \qquad\qquad \alpha\beta = \frac{w_0 w_2}{4 w_1^2} (1 - \alpha - \beta)^2,$$

which is of degree two, and which depends only on the value of $k = \dfrac{w_0 w_2}{4 w_1^2}$.

3) If we change w_1 to $-w_1$, the intrinsic equation does not change; we then find (for $0 \le t \le 1$) another arc $\tilde{\gamma}(t)$ of the same conic (which is no longer in the convex hull of $P_0 P_1 P_2$). If we set

$$\tilde{w}(t) = w_0(1-t)^2 - 2w_1 t(1-t) + w_2 t^2,$$

we have

$$\tilde{\gamma}(t)\tilde{w}(t) - \gamma(t)w(t) = 2w_1 P_1 B_1(t),$$

whence

$$\tilde{\gamma}(t) - P_1 = \frac{w(t)}{\tilde{w}(t)} \left(\gamma(t) - P_1 \right):$$

the points $\gamma(t)$ and $\tilde{\gamma}(t)$ are then on the same line as P_1, and we deduce that the conic C with equation (3.5.8, 1) is the union of the arcs $\gamma(t)$ and

$\tilde{\gamma}(t)$ $(0 \leq t \leq 1)$ (see Figure 3.5.2).

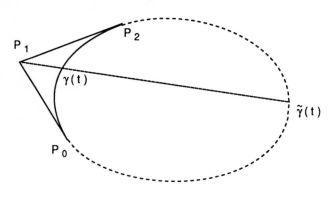

Figure 3.5.2

4) The denominator $\tilde{w}(t)$ of $\tilde{\gamma}(t)$ may have a zero. Its discriminant is equal to $w_1 \sqrt{1 - 4k}$ (with $k = \frac{w_0 w_2}{4 w_1^2}$), which implies

$$\begin{cases} \text{if } 4k > 1 & C \text{ is an ellipse,} \\ \text{if } 4k = 1 & C \text{ is a parabola,} \\ \text{if } 4k < 1 & C \text{ is an hyperbola.} \end{cases}$$

5) We see from the expression of $\gamma(t)$ that $\tilde{\gamma}(t) = \gamma\left(\frac{t}{2t-1}\right)$.

6) Given P_0, P_1 and P_2, the conic C and its parametrization are determined by the data of an additional point P^* and of a value t^* $(0 < t^* < 1)$ such that $\gamma(t^*) = P^*$; this allows to compute $k = \frac{w_0 w_2}{4 w_1^2}$ from the equation (3.5.8, 1).

Examples

a)

$$\begin{cases} P_0 = (1,0) \\ P_1 = (1,1) \\ P_2 = (0,1) \end{cases} \qquad \begin{cases} P^* = (3/5, 4/5) \\ t^* = 1/2 \end{cases}$$

Set $w_0 = 1$: we find then (putting $t = t^*$ in (3.5.7, 1)) $w_1 = 1, w_2 = 2$, whence

$$\begin{cases} X(t) = \dfrac{1 - t^2}{1 + t^2} \\ Y(t) = \dfrac{2t}{1 + t^2}. \end{cases}$$

It is the usual parametrization of the circle S^1 (see Figure 3.5.3).

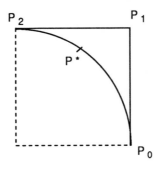

Figure 3.5.3

b) Let C be the circle of center 0 and the points P_0, P_1, P_2 situated as on the Figure 3.5.4 below

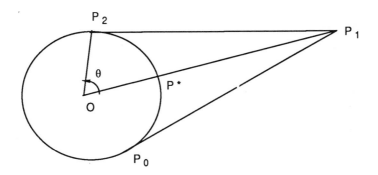

Figure 3.5.4

Let us take $t = 1/2$ and for P^* the middle of the arc P_0P_2; we then find the parametrization

$$\gamma(t) = \frac{P_0(1-t)^2 + 2cos\theta P_1 t(1-t) + P_2 t^2}{(1-t)^2 + 2cos\theta \, t(1-t) + t^2}.$$

7) Let us compute the (symmetry) center S of the conic (in the case

where $k \neq 1/4$, i.e., in the case of an ellipse or an hyperbola).

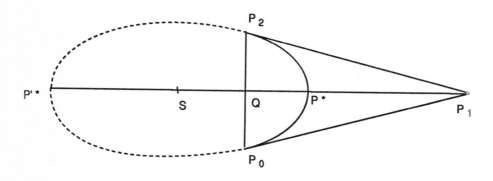

Figure 3.5.5

Let Q be the middle of $P_0 P_2$; the center of C is on the line $P_1 Q$, at the middle of the segment $P^* P'^*$, which is the intersection of $P_1 Q$ with the conic; we find then immediately (with $\gamma(1/2) = P^*$), that

$$ S = \big(2k(P_0 + P_2) - P_1\big)\frac{1}{4k-1}. $$

4

Spline surfaces

4.1 TENSOR PRODUCTS

Let us consider two knot sequences: $u = (u_i)_{0 \le i \le m+k}$, $v = (v_j)_{0 \le j \le n+l}$ satisfying

$$u_0 \le \cdots \le u_k \le a, \qquad u_i \le u_{i+1}$$
$$b \le u_m \le \cdots \le u_{m+k}, \qquad \text{and}$$
$$v_0 \le \cdots \le v_l \le c, \qquad v_j \le v_{j+1}$$
$$d \le v_n \le \cdots \le v_{n+l},$$

and let $B_{i,k}(u)$ (resp. $B_{j,l}(v)$) be the corresponding B-splines.

4.1.1 Definition *With the above notation, let $P_{i,j}$ $(0 \le i \le m - 1, \quad 0 \le j \le n - 1)$ be points in $\mathbf{R}^s$; one then defines a "B-spline patch" as the following application of $[a, b] \times [c, d]$ in $\mathbf{R}^s$*

$$S(u, v) = \sum_{i=0}^{m-1} \sum_{j=0}^{n-1} P_{i,j} B_{i,k}(u) B_{j,l}(v).$$

4.1.2 Remarks

1) We may write

$$S(u, v) = \sum_{i=0}^{m-1} P_i(v) B_{i,k}(u), \quad \text{with}$$

$$P_i(v) = \sum_{j=0}^{n-1} P_{ij} B_{j,l}(v);$$

the image under S of the line segment $v = v_0$ is therefore a spline curve of degree k, with the set of points $P_i(v_0)$ $(0 \le i \le m - 1)$ as control polygon.

We have naturally a similar result for the image under S of the segments $u = \text{constant}$.

2) Each coordinate of the function $S(u, v)$ is of total degree (i.e., in relation to u and v) $l + k$; for instance, in the more common case $l = k = 3$, the coordinates of $S(u, v)$ are of total degree 6.

3) The linear subspace formed with functions on $[a, b] \times [c, d]$ spanned by the $B_{i,k}(u)B_{j,l}(v)$'s $(0 \le i \le m - 1, \ 0 \le j \le n - 1)$ is called the *tensor product* of the spaces $\mathcal{P}_{k,u}$ and $\mathcal{P}_{l,v}$.

4) In the case where

$$\begin{cases} u_0 = \cdots = u_k = a \\ u_m = \cdots = u_{m+k} = b \\ v_0 = \cdots = v_l = c \\ v_n = \cdots = v_{n+l} = d, \end{cases}$$

the boundary of the surface $S(u, v)$ is a union of spline curves which have portions of the boundary of the polyhedron (P_{ij}) as control polygon. In particular, the surface passes through the points $P_{0,0}, \ P_{0,n-0}, \ P_{m-1,n-1}, \ P_{m-1,0}$.

4.1.3 Evaluation algorithm

The advantage of tensor products is that one can use the algorithms developed for the case of one variable. We will see an example of that with the evaluation algorithm. Let

$$S(u, v) = \sum_{i=0}^{m-1} \sum_{j=0}^{n-1} P_{i,j} B_{i,k}(u) B_{j,l}(v)$$

be a B-spline patch, which we want to evaluate at the point (u_0, v_0); let us write as above

$$S(u,v) = \sum_{i=0}^{m-1} P_i(v)B_{i,k}(u), \qquad \text{with}$$

$$P_i(v) = \sum_{j=0}^{n-1} P_{i,j}B_{j,l}(v).$$

We begin by evaluating $P_i(v_0)$ $(0 \leq i \leq m-1)$ by the evaluation algo-rithm 2.3.1 also called "De Boor-Cox" (for each coordinate of $P_i(v_0)$). We then evaluate $S(u_0, v_0)$ using once more this algorithm (for each coordinate of S).

We obtain the following algorithm, assuming

$$(u_0, v_0) \in [u_r, u_{r+1}] \times [v_s, v_{s+1}]$$

(see 2.3.1)

a) **Evaluation of $P_i(v_0)$.** One sets

$$\begin{cases} P_{i,j}^0(v_0) = P_{i,j} \quad (r-k \leq i \leq r,\; s-l \leq j \leq s); \\ P_{i,j}^{p+1}(v_0) = \dfrac{(v_0 - v)P_{i,j}^p(v_0) + (v_{j+l-p} - v_0)P_{i,j-1}^p(v_0)}{v_{j+l-p} - v_j}, \\ \text{for} \quad r-k \leq i \leq r,\; 0 \leq p \leq l-1,\; s-l+p+1 \leq j \leq s; \end{cases}$$

we have $\qquad P_i(v_0) = P_{i,s}^l(v_0).$

b) **Evaluation of $S(u_0, v_0)$ viewed as a spline curve in u with $(P_i(v_0))$ as control polygon.** One sets

$$\begin{cases} P_i^0(u_0, v_0) = P_{i,s}^l(v_0) \quad \text{for} \quad r-k+1 \leq i \leq r; \\ P_i^{q+1}(u_0, v_0) = \dfrac{(u_0 - u_i)P_i^q(u_0, v_0) + (u_{i+k-q} - u_0)P_{i-1}^q(u_0, v_0)}{u_{i+k-q} - u_i}, \\ \text{for} \quad 0 \leq q \leq k-1 \quad \text{and} \quad r-k+q+1 \leq i \leq r; \end{cases}$$

we have at last $\qquad S^r(u_0, v_0) = P_r^k(u_0, v_0).$

There is clearly a similar algorithm obtained by exchanging the roles of u and v.

4.1.4 Example   In the bicubic case $(k = l = 3)$, we obtain the following figure in $\mathbf{R}^3$

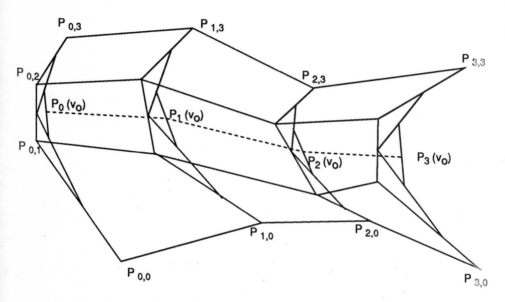

Figure 4.1.1 : Computation of $P_i(v_0)$ ($v_3 < v_0 < v_4$)

4.1.5   In the same manner, we may apply the subdivision algorithm 2.3.19 to the tensor product, assuming that the knots are points of $\mathbf{Z} \times \mathbf{Z}$. Set

$$S(X,Y) = \sum_{i,j \in \mathbf{Z}^2} P_{i,j} B_k(X - i) B_k(Y - j)$$

where $B_k(X)$ is the B-spline function of degree k with knots $(0, 1, \ldots, k + 1)$ (see Chapter 2, Section 2.3). If then we set as in Chapter 2, Section 2.3

$$B_k(X) = \sum_{i=0}^{(k+1)(p-1)} a_p^k(i/p) B_k(pX - i),$$

we obtain

$$S(X,Y) = \sum_{m,n \in \mathbf{Z}^2} Q_p^k\left(\frac{m}{p}, \frac{n}{p}\right) B_k(pX - m) B_k(pY - n)$$

with

$$Q_p^k\left(\frac{m}{p},\frac{n}{p}\right) = \sum_{i,j\in\mathbf{Z}^2} a_p^k\left(\frac{m}{p}-i\right)a_p^k\left(\frac{n}{p}-j\right)P_{i,j}$$

4.2 PARTICULAR CASE OF BEZIER SURFACES

Let us take $(u,v) \in [0,1]^2$, and consider

$$B(u,v) = \sum_{i=0}^{k}\sum_{j=0}^{l} P_{i,j}B_i^k(u)B_j^l(v),$$

where $B_i^k(u)$ is the ith Bernstein polynomial of degree k (2.1.2). The image of $B(u,v)$ is generally called a "Bézier patch".

4.2.1 Computation of the derivatives

Let D^{rs} denote the differential operator $\left(\frac{\partial}{\partial u}\right)^r\left(\frac{\partial}{\partial v}\right)^s$; we have

$$D^{rs}B(u,v) = \frac{k!}{(k-r)!}\frac{l!}{(l-s)!}\sum_{i=0}^{k-r}\sum_{j=0}^{l-s} \Delta^{rs}P_{i,j}B_i^{k-r}(u)B_j^{l-s}(v),$$

with

$$\Delta^{rs}P_{i,j} = \Delta_1^r\big(\Delta_2^s P_{i,j}\big) = \Delta_2^s\big(\Delta_1^r P_{i,j}\big),$$

where

$$\begin{cases} \Delta_1 P_{i,j} = P_{i+1,j} - P_{i,j} \\ \Delta_2 P_{i,j} = P_{i,j+1} - P_{i,j} \end{cases}$$

(see 2.3.5).

This computation permits the junction of several Bézier patches. Let us for instance consider two Bézier patches of the same degree (k,l): $B(u,v)$ $((u,v) \in [0,1]^2)$ with $(P_{i,j})$ as control polygon, and $B'(s,t)$ $((s,t) \in [0,1]^2)$ with $P'_{i,j}$ as control polygon.

Assume that the junction is made along the curves $B(1,v)$ and $B'(0,t)$

$$\begin{cases} B(1,v) = \sum_{j=0}^{l} P_{k,j}B_j^l(v) \\ B'(0,t) = \sum_{j=0}^{l} P'_{0,j}B_j^l(t). \end{cases}$$

a) If we want $\mathcal{C}^0$ continuity between these two patches, the two curves have to coincide, i.e., we must have $P_{k,j} = P'_{0,j}$ $(0 \le j \le l)$.

b) Let us compute the first partial derivatives of B and B'

$$\begin{cases} \dfrac{\partial B}{\partial u}(1,v) = D_{1,0}B(1,v) = k\displaystyle\sum_{j=0}^{l}(P_{k,j} - P_{k-1,j})B_j^l(v) \\[4mm] \dfrac{\partial B}{\partial v}(1,v) = D_{0,1}B(1,v) = l\displaystyle\sum_{j=0}^{l-1}(P_{k,j+1} - P_{k,j})B_j^{l-1}(v), \end{cases}$$

and in the same manner

$$\begin{cases} \dfrac{\partial B'}{\partial s}(0,t) = k\displaystyle\sum_{j=0}^{l}(P'_{1,j} - P'_{0,j})B_j^l(t) \\[4mm] \dfrac{\partial B'}{\partial t}(0,t) = l\displaystyle\sum_{j=0}^{l-1}(P'_{0,j+1} - P'_{0,j})B_j^{l-1}(t); \end{cases}$$

note that we have $\frac{\partial B}{\partial v}(1,v) = \frac{\partial B'}{\partial t}(0,v)$ $(v \in [0,1])$ if condition a) is fulfilled.

If we want $\mathcal{C}^1$ continuity, it is necessary that $P_{k-1,j} + P'_{1,j} = 2P_{k,j}$ $(0 \le j \le l)$. In other words, the vertices $P_{k,j} = P'_{0,j}$ have to be at the middle of the segments $P_{k-1,j}P'_{1,j}$, which expresses that for each j, the Bézier curves (in u) with control polygons $P_{i,j}$ $(0 \le i \le k)$ and $P'_{i,j}$ $(0 \le i \le k)$ have $\mathcal{C}^1$ continuity at $P_{k,j} = P'_{0,j}$. (see Chapter 3, Section 3.4, b)).

4.2.2 Remarks

1) In the same manner, we can give conditions for $\mathcal{C}^r$ continuity $(r > 1)$ for the two Bézier patches. It is for instance easy to see that, as well as conditions studied in a) and b), for $\mathcal{C}^2$ continuity along the curves $B(1,v)$ and $B'(0,t)$, the following must be fulfilled:

$$P'_{2,j} - P_{k-2,j} = 2(P'_{1,j} - P_{k-1,j}) \qquad (0 \le j \le l)$$

(see Chapter 3, Section 3.4, b)). More generally, we immediately see that two patches $B(u,v)$ and $B'(s,t)$ have a $\mathcal{C}^r$ continuity along the curves $B(1,v)$ and $B'(0,t)$ if and only if, for $0 \le j \le l$, the Bézier curves with control polygons $P_{i,j}$ and $P'_{i,j}$ $(0 \le i \le k)$ have a $\mathcal{C}^r$ continuity at $P_{k,j} = P'_{0,j}$.

2) The $\mathcal{C}^r$ continuity condition may be interpreted in the following way. Set $s = u - 1$, $t = v$; then the surface $B(u,v)$ $0 \leq u \leq 2$, $0 \leq v \leq 1$, defined by $B(u,v) = B(u,v)$ for $0 \leq u \leq 1$ and $B(u,v) = B'(u,v)$ for $1 \leq u \leq 2$ is of class $\mathcal{C}^r$.

3) If the patches B and B' are parametrized by $(u_1, v_1) \in [a, b] \times [c, d]$ and $(s_1, t_1) \in [b, b'] \times [d, d']$, and if we want $\mathcal{C}^r$ continuity with respect to the parameters (u_1, v_1) and (s_1, t_1), we come down to the preceding case, setting

$$u = \frac{u_1 - a}{b - 1} \qquad v = \frac{v_1 - c}{d - c}$$

$$s = \frac{s_1 - b}{b' - b} \qquad t = \frac{t_1 - d}{d' - d}.$$

The $\mathcal{C}^1$ continuity is then expressed by the relations

$$(b' - b)P_{k-1,j} + (b - a)P'_{1,j} = (b' - a)P_{k,j} \quad (0 \leq j \leq l).$$

The points $P_{k-1,j}$, $P_{k,j} = P'_{0,j}$ and $P'_{1,j}$ are again on the same line, but $P_{k,j}$ is no longer necessarily at the middle of $[P_{k-1,j}, P'_{1,j}]$.

4) The condition $\lambda P_{k-1,j} + \mu P'_{1,j} = (\lambda + \mu)P_{k,j}$ (see 2) above) says that the two Bézier patches have the same tangent plane at the point $P_{k,j} = P'_{0,j}$. More generally, one may define the condition for G_i continuity ($i \geq 0$) between two parametric surfaces, in a similar way as for curves (see 3.4.5).

4.2.3 Definition   One says that two parametric surfaces $S_1(u,v)$ $(0 \leq u \leq 1,\ 0 \leq v \leq 1)$ and $S_2(s,t)$ $(0 \leq s \leq 1,\ 0 \leq t \leq 1)$ have G_i continuity $(i \geq 0)$ along the curves $S_1(1,v)$ and $S_2(0,t)$ (we have $S_1(1,v) = S_2(0,t)$ $0 \leq v \leq 1,\ 0 \leq t \leq 1)$ by assumption), if there exists a change of parameters

$$(s', t') \longmapsto (s, t) = \big(\psi_1(s', t'),\ \psi_2(s', t')\big),$$

$$with \quad \psi_1(0, t') = 0,\ \psi_2(0, t') = t',$$

such that the surfaces $S_1(u,v)$ and $S_2\big(\psi_1(s', t'),\ \psi(s', t')\big)$ have C^i continuity along the curves $S_1(1, v)$ and $S_2\big(\psi_1(0, t'),\ \psi_2(0, t')\big) = S_2(0, t')$.

As with curves, the condition of G_0 continuity is the same as the C^0 one. Let us look to the G_1 continuity condition.

4.2.4 Lemma   *Assume that there is a C^0 junction between the two surfaces $S_1(u,v)$ and $S_2(s,t)$ along the curve γ : $S_1(0,v) = S_2(1,t)$.*

There is G_1 continuity along this curve if and only if S_1 and S_2 have the same tangent plane at all the points of the common curve γ.

Proof It is clear that if S_1 and S_2 have G_1 continuity along γ, they have the same tangent plane, as the tangent plane to a surface is independent of the choosen parametrization.

Conversely, assume that the two surfaces have the same tangent plane at each point of γ. The claim follows then easily from the implicit function theorem: let us look at a point $O \in \gamma$ which we assume to be the origin of $\mathbf{R}^3$, with coordinates X, Y, Z. We can then assume, from the implicit function theorem, that the surface S_1 is defined by the equation $Z = 0$ in a neighborhood of O, and we can take X and Y as parameters for S_2, which is then defined by an equation of the form $Z = \phi(X, Y)$, with $\phi(0,0) = 0$, and $\partial\phi/\partial X(0,0) = \partial\phi/\partial Y(0,0) = 0$ by assumption.

The surfaces S_1 and S_2, with parameters X and Y, are now glued at O in a surface of class C^1.

∎

4.2.5   Let us apply the above fact to the case of two Bézier patches $B(u,v)$ and $B'(u,v)$, gluing each other along the curve $B(1,v) = B'(0,v)$ (one then has $P_{k,j} = P'_{0,j}$ $(0 \le j \le l)$).

The tangent plane continuity along the common curve is expressed by the following relation (see 4.2.1)

$$\alpha(v)k \sum_{j=0}^{l}(P_{k,j} - P_{k-1,j})B_j^l(v) + \beta(v)k \sum_{j=0}^{l}(P'_{1,j} - P'_{0,j})B_j^l(v)$$

$$+\gamma(v)l \sum_{j=0}^{l-1}(P_{k,j+1} - P_{k,j})B_j^{l-1}(v) = 0,$$

$\alpha(v)$, $\beta(v)$ and $\gamma(v)$ being functions (a priori of any kind), of v.

It is natural to assume that α, β and γ are polynomials in v; the simplest choice is to take α and β constants, and γ linear in v, which lets us use the linear independence of the $B_j^l(v)$'s; one then sets

$$\begin{cases} \alpha(v) = \alpha \\ \beta(v) = \beta \\ \gamma(v) = \gamma_0(1-v) + \gamma_1(v). \end{cases}$$

Since

$$\begin{cases} (1-v)B_j^{l-1}(v) = \dfrac{l-j}{l}B_j^l(v) & (0 \le j \le l-1) \\[2mm] vB_j^{l-1}(v) = \dfrac{j+1}{l}B_{j+1}^l(v) & (0 \le j \le l-1), \end{cases}$$

(see 2.1.2), one finds the following relations

$$\begin{cases} P_{k,j} = P'_{0,j}, \\ \alpha k(P_{k,j} - P_{k-1,j}) + \beta k(P'_{1,j} - P'_{0,j}) \\ \quad + \gamma_0(l-j)(P_{k,j+1} - P_{k,j}) + \gamma_1(j)(P_{k,j} - P_{k,j-1}) = 0 & (0 \le j \le l), \end{cases}$$

which give sufficient conditions for the G_1 continuity of two Bézier patches along their common boundary $B(1,v) = B'(0,v)$.

For more details about these matters, the reader may look at [Sc-Du] or [Du].

4.2.6 Matrix representation

Let us consider a bicubic patch (i.e., $k = l = 3$). Recall (Chapter 3, Section 3.3, b)) that a cubic Bézier curve is represented in the following way

$$f(t) = \begin{pmatrix} 1 & t & t^2 & t^3 \end{pmatrix} \begin{pmatrix} 1 & 0 & 0 & 0 \\ -3 & 3 & 0 & 0 \\ 3 & -6 & 3 & 0 \\ -1 & 3 & -3 & 1 \end{pmatrix} \begin{pmatrix} P_0 \\ P_1 \\ P_2 \\ P_3 \end{pmatrix}$$

4.2.7 Proposition   Let

$$P(u,v) = \sum_{i=0}^{3}\sum_{j=0}^{3} P_{i,j}B_i^3(u)B_j^3(v)$$

be a Bézier patch, with $P_{i,j} \in \mathbf{R}^s$; one may then write

$$P(u,v) = UMP\,{}^tM\,{}^tV, \quad with$$

$U = \begin{pmatrix} 1 & u & u^2 & u^3 \end{pmatrix}$, $V = \begin{pmatrix} 1 & v & v^2 & v^3 \end{pmatrix}$, $P = (P_{i,j})_{0 \le i \le 3,\, 0 \le j \le 3}$, and

$$M = \begin{pmatrix} 1 & 0 & 0 & 0 \\ -3 & 3 & 0 & 0 \\ 3 & -6 & 3 & 0 \\ -1 & 3 & -3 & 1 \end{pmatrix}.$$

Proof One may write

$$P(u,v) = \Big(B_0(u),\dots,B_3(u)\Big) P \begin{pmatrix} B_0(v) \\ \vdots \\ B_3(v) \end{pmatrix};$$

it is now enough to apply the following formula

$$\Big(B_0(u)\dots B_3(u)\Big) = \Big(1 \quad u \quad u^2 \quad u^3\Big) M$$

and the similar formula obtained by exchanging u and v, then transposing.

∎

4.3 INTERPOLATION AND APPROXIMATION

a) Interpolation of a set of points

Let $Q_{\lambda,\mu}$ $(0 \le \lambda \le m-1,\ 0 \le \mu \le n-1)$ be a set of points in $\mathbf{R}^s$. The interpolation problem consists of finding a spline surface

$$S(u,v) = \sum_{i=0}^{m-1}\sum_{j=0}^{n-1} P_{i,j} B_{i,k}(u) B_{j,l}(v)$$

and parameters $(\alpha_\lambda, \beta_\mu)$ such that

(4.3.1) $S(\alpha_\lambda,\beta_\mu) = Q_{\lambda,\mu}$ $(0 \le \lambda \le m-1,\ 0 \le \mu \le n-1).$

As in Chapter 3, Section 1, we will assume that $B_{\lambda,k}(\alpha_\lambda)B_{\mu,l}(\beta_\mu) \ne 0$ for all the pairs (λ,μ); this is equivalent, by 3.1.1, to assuming that the matrices

$$\begin{cases} A = (a_{\lambda,i}) = \Big(B_{i,k}(\alpha_\lambda)\Big) & (\lambda,i) \in [0,m-1] \times [0,m-1] \\ B = (b_{\mu,j}) = \Big(B_{j,l}(\beta_\mu)\Big) & (\mu,j) \in [0,n-1] \times [0,n-1] \end{cases}$$

are *invertible*.

(4.3.1) then becomes: $Q = A P\,^t B$ where Q is the matrix $(Q_{i,j})$ and P the matrix $(P_{i,j})$ (which we have to determine). To solve this equation, one sets $U = P\,^t B$, and computes U solving $AU = Q$ (if the points $Q_{\lambda,\mu}$ are in $\mathbf{R}^3$, this requires us solving $3n$ linear systems with matrix A: for that, see the remarks of Chapter 3, Section 3.1).

One then solves the system $B^t P = {}^t U$ (to determine the matrix P), which is equivalent to $3m$ scalar linear systems with matrix B.

b) Approximation of a surface (or "quasi-interpolation")

If now we want *approximate* a parametric surface generated by the function $f(u,v)$ defined on $[a,b] \times [c,d]$, we may, as in Chapter 1, Section 1.6, set

$$S_f(u,v) = \sum_{i=0}^{m-1} \sum_{j=0}^{n-1} f(\theta_i, \eta_j) B_{i,k}(u) B_{j,l}(v),$$

with

$$\begin{cases} \theta_i = \dfrac{u_{i+1} + \cdots + u_{i+k}}{k} \\[2mm] \eta_j = \dfrac{v_{j+1} + \cdots + v_{j+l}}{l}. \end{cases}$$

With this choice of θ_i and η_j, this spline surface will have the property, as in 1.6.4, *to soften the oscillations* of the surface defined by f.

c) Boolean sum of two operators

To approximate the surface, one may also make the *boolean sum* of the two operators (with the above notation):

$$\begin{cases} S_m g(u) = \sum_{i=0}^{m-1} g(\theta_i) B_{i,k}(u) \\[2mm] S_n h(v) = \sum_{i=0}^{n-1} h(\eta_j) B_{j,l}(v), \end{cases}$$

setting

$$(S_m \oplus S_n) f(u,v) = S_m f(.,v) + S_n f(u,.) - S_f(u,v)$$

$$= \sum_{i=0}^{m-1} f(\theta_i, v) B_{i,k}(u) + \sum_{j=0}^{n-1} f(u, \eta_j) B_{j,l}(v)$$

$$- \sum_{i=0}^{m-1} \sum_{j=0}^{n-1} f(\theta_i, \eta_j) B_{i,k}(u) B_{j,l}(v).$$

As $S_m \ell = \ell$ (and $S_n \ell = \ell$) for any linear form ℓ (see 1.6.1), we see that

$$\begin{cases} (S_m \oplus S_n)\ell(u)v^j = \ell(u)v^j & (0 \le j \le n-1) \\ (S_m \oplus S_n)u^i \ell(v) = u^i \ell(v) & (0 \le i \le m-1) \end{cases}$$

for any linear form ℓ; in particular, $(S_m \oplus S_n)f(u, v) = f(u, v)$ for any polynomial P of degree ≤ 3. This operator gives then a very good approximation of the surface, but the operation needs the knowledge of the "level lines" $f(u, \eta_j)$ and $f(\theta_i, v)$.

d) Coons patches

One may also consider the linear operator

$$r_1 g(u) = \frac{1}{b-a}\left[(b-u)g(a) + (u-a)g(b)\right]$$

which, for a function g of one variable defined on an interval $[a, b]$, interpolates $g(a)$ and $g(b)$. Set in the same way

$$r_2 h(v) = \frac{1}{d-c}\left[(d-v)h'c) + (v-c)h(d)\right]$$

which interpolates $h(c)$ and $h(d)$ for a function h defined on an interval $[c, d]$.

For a function $f(u, v)$ $\Big((u, v) \in [a, b] \times [c, d]\Big)$, the direct sum $(r_1 \oplus r_2)f(u, v)$ becomes

$$\frac{1}{b-a}\left[(b-u)f(a, v) + (u-a)f(b, v)\right]$$
$$+ \frac{1}{d-c}\left[(d-v)f(u, c) + (v-c)f(u, d)\right]$$
$$- \frac{1}{(b-a)(d-c)}\Big[(b-u)(d-v)f(a, c) + (b-u)(v-c)f(a, d)$$
$$+ (u-a)(d-v)f(b, c) + (u-a)(v-c)f(b, d)\Big].$$

This surface, called a "Coons patch", interpolates the four curves of the boundary of the parametric surface defined by $f(u, v)$, namely the curves $f(u, c)$, $f(u, d)$, $f(a, v)$, $f(b, v)\big((u, v) \in [a, b] \times [c, d]\big)$. Its construction needs in fact only the knowledge of these four curves.

4.4 BERNSTEIN POLYNOMIALS

Tensor products are often used because they are easy to compute, and because one can apply to them the algorithms of the one variable case; nevertheless, they have several disadvantages:

a) They are well adapted for the rectangular portions of surfaces, but not for triangular ones for instance.

b) There are two privileged families of curves on a spline surface defined as a tensor product (the ones defined by $u = $ constant, and by $v = $ constant); one can, for these curves, which are one variable splines, easily get many properties (for instance, curvature, variation diminishing, etc ...) On the other hand, it is very difficult to study curves in other directions, for instance $u + v = $ constant .

c) Even in the bicubic case, the most often used, each coordinate is of *total degree 6* in u and v, which implies that the intrinsic equation $f(X, Y, Z) = 0$ of a bicubic patch is a portion of an algebraic surface of degree 18. The intersection of two such surfaces is then an algebraic curve of degree 324, hardly accessible to formal computation.

The above considerations lead us to try to generalize directly spline curves (and Bézier curves) to the case of surfaces. We will begin with the case of Bézier patches, and then study, briefly, the case of polyhedral splines.

Let $\Delta \subset \mathbf{R}^2$ be a triangle with vertices a, b, c and of area 1, and let (r, s, t) be the barycentric coordinates of a point $P \in \mathbf{R}^2$ with respect to (a, b, c).

Recall that

- a) $r + s + t = 1$.
- b) P is in the interior of the triangle Δ if and only if $r > 0, s > 0, t > 0$.
- c) r, s and t are respectively the areas of the triangles Pbc, Pac

and *Pab* (Figure 4.4.1).

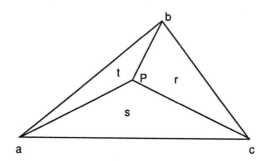

Figure 4.4.1

4.4.1 Definition   *Let us fix a triangle Δ and an integer n; the Bernstein polynomials of degree n are defined in the following way*

$$B^n_{i,j,k} = \frac{n!}{i!j!k!} r^i s^j t^k \quad \text{for} \quad i+j+k = n.$$

4.4.2 Properties of the $B_{i,j,k}$'s

1) There are $\dfrac{(n+1)(n+2)}{2}$ Bernstein polynomials of degree $\leq n$.

2) The $B^n_{i,j,k}(r, s, 1-r-s)$'s form a *basis* of the space of polynomials in r and s of degree $\leq n$; this basis is called the *Bernstein basis*.

For the proof, it is enough, after 1), to show that the $B^n_{i,j,k}(r, s, 1-r-s)$'s are linearly independent, which is clear, as the lowest degree term in the expression $r^i s^j (1-r-s)^k$ is $r^i s^j$.

3) One has the following induction formula

$$B^n_{i,j,k} = r B^{n-1}_{i-1,j,k} + s B^{n-1}_{i,j-1,k} + t B^{n-1}_{i,j,k-1}.$$

In fact, the coefficient of the term $r^i s^j t^k$ on the right is

$$\frac{(n-1)!}{(i-1)!(j-1)!(k-1)!} \left[\frac{1}{jk} + \frac{1}{ik} + \frac{1}{ij} \right],$$

which is equal to $\frac{n!}{i!j!k!}$, as $i+j+k = n$.

4) One has

$$\sum_{i+j+k=n} B^n_{i,j,k}(r,s,1-r-s) = 1.$$

In fact, $\sum_{i+j+k=n} B^n_{i,j,k}(r,s,t)$ is equal to the development of $(r+s+t)^n$; we have then only to set $t = 1-r-s$.

5) If $P \in \mathbf{R}^2$ has barycentric coordinates r,s,t with respect to the triangle Δ, one has $B^n_{i,j,k}(r,s,t) > 0$ in the interior of Δ.

6) *Degree elevation*

Any polynomial of degree n (in r,s,t) may be viewed as a polynomial of degree $\leq n+1$, and so expanded in the basis $B^{n+1}_{i,j,k}$. One has in particular $B^n_{i,j,k} = (r+s+t)B^n_{i,j,k}(r,s,t)$, because $r+s+t = 1$, and then

$$B^n_{i,j,k} = \frac{1}{n+1}\left[(i+1)B^{n+1}_{i+1,j,k} + (j+1)B^{n+1}_{i,j+1,k} + (k+1)B^{n+1}_{i,j,k+1}\right],$$

as one immediately sees, after recalling Definition 4.4.1.

4.5 "TRIANGULAR" BEZIER PATCHES

In a way similar to the study of curves (see 2.2.1), let us fix an integer n, a triangle $\Delta \subset \mathbf{R}^2$, and let us consider a "triangular" network $(P_{i,j,k})$ $(i+j+k = n)$ of points in $\mathbf{R}^s$.

4.5.1 Definition *A Bézier triangular patch is by definition the parametric surface generated by the map $B : \overline{\Delta} \longrightarrow \mathbf{R}^s$, defined as follows*

$$B(r,s,t) = \sum_{i+j+k=n} P_{i,j,k} B^n_{i,j,k}(r,s,t)$$

($\overline{\Delta}$ denotes the closure of the interior of the triangle Δ.)

4.5.2 Example: case n = 3

The basis $B^n_{i,j,k}(r,s,t)$ is made from 10 elements, which we may represent in the following triangular form

$$
\begin{array}{ccccccc}
 & & & s^3 & & & \\
 & & 3s^2t & & 3rs^2 & & \\
 & 3st^2 & & 6rst & & 3r^2s & \\
t^3 & & 3rt^2 & & 3r^2t & & r^3
\end{array}
$$

Let us draw the points $P_{i,j,k}$ in $\mathbf{R}^3$, assuming that their disposition reflects that of the above triangle (one may for instance assume that if π denotes a linear projection: $\mathbf{R}^3 \longrightarrow \mathbf{R}^2$, $\pi(P_{i,j,k})$ is the point of barycentric coordinates $\left(\frac{i}{n}, \frac{j}{n}, \frac{k}{n}\right)$).

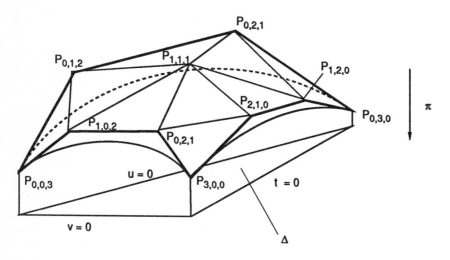

Figure 4.5.1

Note that the points $P_{i,j,k}$ have been joined in such a way that they form a polyhedron (called "control polyhedron") having triangles as faces.

The choice of these triangles is not unique (the $P_{i,j,k}$'s being given); the way chosen here gives a polyhedron only if the points $P_{i,j,k}$ have a disposition as explained above (as for instance in Figure 4.5.1, where the point $\pi(P_{i,j,k})$ has barycentric coordinates $\left(\frac{i}{n}, \frac{j}{n}, \frac{k}{n}\right)$); this choice corresponds to the *Delaunay triangulation* of the triangle Δ, with vertices $\pi(P_{i,j,k})$ when they have barycentric coordinates $\left(\frac{i}{n}, \frac{j}{n}, \frac{k}{n}\right)$ (see Chapter 5).

Let us for instance look to the case $n = 3$, with a triangle Δ in $\mathbf{R}^2$, and the vertices $Q_{i,j,k} = \pi(P_{i,j,k})$ $(i+j+k = 3)$ whose barycentric coordinates are $\left(\frac{i}{3}, \frac{j}{3}, \frac{k}{3}\right)$ (see Figure 4.5.2).

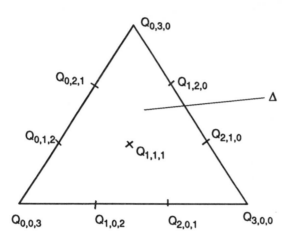

Figure 4.5.2

One may think of several triangulations of Δ (i.e., decompositions in triangles having the points $Q_{i,j,k}$ as vertices), for instance the two triangulations
of Figure 4.5.3.

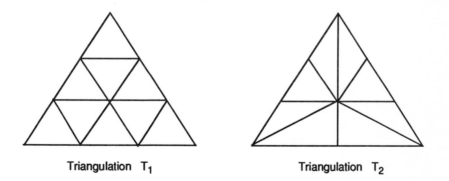

Triangulation T_1 Triangulation T_2

Figure 4.5.3

The triangulation T_1 is called a " Delaunay triangulation (or tesselation)",
and its properties will be studied (in the general case) in Chapter 5. We
will systematically adopt it; one sees in particular in Figure 4.5.3 that it
produces triangles with less acute angles than the triangulation T_2.

4.5.3 Properties of the patch $B(r, s, t)$

1) The patch $B(r, s, t) = \sum_{i+j+k=n} P_{i,j,k} B^n_{i,j,k}(r, s, t)$ passes through the points $P_{n,0,0}$, $P_{0,n,0}$ and $P_{0,0,n}$; one has in fact

$$B^n_{n,0,0}(1, 0, 0) = B^n_{0,n,0}(0, 1, 0) = B^n_{0,0,n}(0, 0, 1) = 1.$$

2) The boundary of the patch is formed with three Bézier curves, images of the sides of the triangle Δ (these sides are defined by $r = 0$, $s = 0$, and $t = 0$).

For instance, the curve image of the side $r = 0$ is the Bézier curve of degree n with control polygon $(P_{0,0,n}, P_{0,1,n-1}, \ldots, P_{0,n,0})$; the $B^n_{0,j,l}(0, s, 1 - s)$'s are in fact the Bernstein polynomials of degree n in s, as it is immediately seen.

3) The patch is included in the convex hull of the points $P_{i,j,k}$. This results from the relation $\sum_{i+j+k=n} B^n_{i,j,k} = 1$ (see 4.4.2, 4)).

4) The tangent planes to the patch $B(r, s, t)$ at the points $P_{n,0,0}$, $P_{0,n,0}$, $P_{0,0,n}$ are generated by the two neighbouring points of the network; for instance, the tangent plane to the patch at the point $P_{n,0,0}$ is the plane spanned by the points $(P_{n,0,0}, P_{n-1,1,0}, P_{n_1,0,1})$.

This results immediately from the fact that the vectors $\overrightarrow{P_{n,0,0}P_{n-1,1,0}}$ and $\overrightarrow{P_{n,0,0}P_{n-1,0,1}}$ are tangent to the two Bézier curves of the boundary of the patch, passing through the point $P_{n,0,0}$.

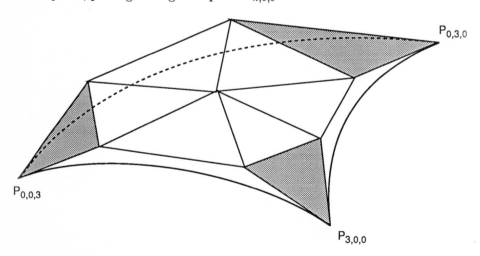

Figure 4.5.4 : The tangent planes are shaded

5) The image under B of any straight line of the r, s, t plane, is a parametric curve of degree n. (Note that the similar property for tensor products is false.)

Let us now look at some algorithms useful for computing Bézier patches.

4.5.4 Evaluation algorithm (or "De Casteljau algorithm")

It simply uses the induction formula given in 4.4.2, 3).

Let $B(r, s, t) = \sum P_{i,j,k} B^n_{i,j,k}(r, s, t)$ be a triangular Bézier patch; we want to evaluate $B(r, s, t)$ for fixed values of r, s, t.

a) One sets $P^0_{i,j,k} = P_{i,j,k}$ $(i + j + k = n)$.

b) For $1 \le \ell \le n$, solve

$$P^\ell_{i,j,k} = r P^{\ell-1}_{i+1,j,k} + s P^{\ell-1}_{i,j+1,k} + t P^{\ell-1}_{i,j,k+1} \quad \text{for } i + j + k = n - \ell.$$

c) We have $B(r, s, t) = P^n_{0,0,0}(r, s, t)$.

Let us illustrate this algorithm in the case of a patch in $\mathbf{R}^3$ of degree 3, whose image under the projection π is the triangle Δ: let $P_{i,j,k}$ $(i + j + k = 3)$ be points in $\mathbf{R}^3$ such that $\pi(P_{i,j,k}) = Q_{i,j,k} = (\frac{i}{3}, \frac{j}{3}, \frac{k}{3}) \in \Delta$, and set $Q^\ell_{i,j,k} = \pi(P^\ell_{i,j,k})$ $(0 \le \ell \le 3)$.

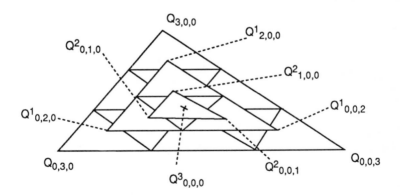

Figure 4.5.5

Figure 4.5.5 shows this evaluation algorithm; one has $B(r, s, t) = Q^3_{0,0,0}$.

4.5.5 Remark
The De Casteljau algorithm above, which has a rather high cost in number of operations, may give more information than the simple evaluation of $B(r, s, t)$.

We will now see for instance that the intermediate points computed during the De Casteljau algorithm will be useful for the computation of the derivatives; in particular, the tangent plane to the Bézier patch at the point $B(r,s,t)$ is the plane containing the points $P_{1,0,0}^{n-1}$, $P_{0,1,0}^{n-1}$ and $P_{0,0,1}^{n-1}$. Moreover, the De Casteljau algorithm also gives, as for Bézier curves, a subdivision algorithm of the Bézier patch (this time in three patches).

Computation of the derivatives

Let us keep the notation of 4.5.1, and let us consider a line in the plane with barycentric coordinates (r,s,t), parametrized by: $X(\tau) = X_0(1-\tau) + X_1\tau$, with $X_0 = (r_0,s_0,t_0)$ and $X_1 = (r_1,s_1,t_1)$. We will compute the derivative of $B\big(X(\tau)\big)$, i.e., the differential operator D_X of B, in the direction of the curve $B\big(X(\tau)\big)$.

4.5.6 Proposition   *We have*

$$D_X B(X_0) =$$
$$n \sum_{i+j+k=n-1} B_{i,j,k}^{n-1}(r_0,s_0,t_0)\Big(\Delta r_0 P_{i+1,j,k} + \Delta s_0 P_{i,j+1,k} + \Delta t_0 P_{i,j,k+1}\Big),$$

with

$$\begin{cases} \Delta r_0 = r_1 - r_0 \\ \Delta s_0 = s_1 - s_0 \\ \Delta t_0 = t_1 - t_0. \end{cases}$$

Proof   The formula is immediate, differentiating

$$B_{i,j,k}^n(\tau) = \frac{n!}{i!j!k!}r^i(\tau)s^j(\tau)t^k(\tau)$$

with respect to τ, and setting $\tau = 0$ in the result.

∎

4.5.7 Corollary   *The tangent plane to the patch B (at the point $B(r,s,t)$) is the plane defined by the points $P_{1,0,0}^{n-1}$, $P_{0,1,0}^{n-1}$, and $P_{0,0,1}^{n-1}$, obtained in the De Casteljau algorithm.*

Proof   We have, by Proposition 4.5.6,

$$D_X B(X_0) = n\Big[\Delta r_0 \sum B_{i,j,k}^{n-1}(r_0,s_0,t_0)P_{i+1,j,k}$$
$$+\Delta s_0 \sum B_{i,j,k}^{n-1}P_{i,j+1,k}(r_0,s_0,t_0) + \Delta t_0 \sum B_{i,j,k}^{n-1}(r_0,s_0,t_0)P_{i,j,k+1}\Big]$$

the sums being taken for $i + j + k = n - 1$.

We may now apply the De Casteljau algorithm 4.5.4 to the three above sums (for the fixed values $(r, s, t) = (r_0, s_0, t_0)$ of the parameters; but for instance the first sum

$$\sum_{i+j+k=n-1} B_{i,j,k}^{n-1} P_{i+1,j,k}$$

is a triangular Bézier patch of degree $n - 1$, constructed with the network of points $\left(P_{i,j,k}\right)_{(i \geq 1)}$, and the algorithm gives the same computations than 4.5.4 for these points, finishing at the point $P_{1,0,0}^{n-1}$; we then finally obtain

$$D_X B(X_0) = n\left[\Delta r_0 P_{1,0,0}^{n-1} + \Delta s_0 P_{0,1,0}^{n-1} + \Delta t_0 P_{0,0,1}^{n-1}\right],$$

which implies that the tangent line at $B(r_0, s_0, t_0)$ to the curve $B(X(\tau))$ is contained in the plane defined by $P_{1,0,0}^{n-1}$, $P_{0,1,0}^{n-1}$ and $P_{0,,0,1}^{n-1}$. ∎

4.5.8 Remark   One may generalize the formula in 4.5.6 to higher order derivatives, which lets us calculate the Taylor series, i.e., the polynomial expression of $B(r, s, t)$ at a given point.

4.5.9 Degree elevation

Let us look at a patch

$$B(r, s, t) = \sum_{i+j+k=n} P_{i,j,k} B_{i,j,k}^{n}.$$

Applying 4.4.2, 6), we obtain

$$B(r, s, t) = \sum_{i+j+k=n+1} P_{i,j,k}^{*} B_{i,j,k}^{n+1},$$

with a new network $\left(P_{i,j,k}^{*}\right)_{i+j+k=n+1}$ defined by

$$P_{i,j,k}^{*} = \frac{1}{n+1}\left(i P_{i-1,j,k} + j P_{i,j-1,k} + k P_{i,j,k-1}\right).$$

One may easily prove, as for curves, that if one does this operation several times, the sequence of obtained networks is converging towards the patch.

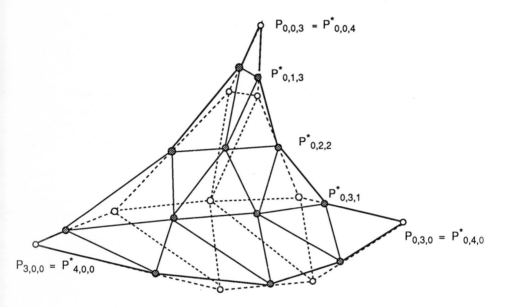

Figure 4.5.7 : Degree elevation from 3 to 4

4.6 JUNCTION BETWEEN BEZIER PATCHES

To be able to model surfaces, it is necessary to be able to have several triangular patches at a junction.

C^1 junction of two triangular patches

Let Δ and Δ' be two triangles, with barycentric coordinates respectively (r, s, t) and (r', s', t'), and two Bézier patches of the same degree n, $B(r, s, t)$ and $B'(r', s', t')$ (one may assume that the two patches have the same degree because of 4.5.9).

We may make the junction of the two patches for instance along the curves $r = 0$ and $r' = 0$. To have a continuity in the parametrization, we will then assume that Δ has the side $r = 0$ common with the side $r' = 0$ of

Δ', Δ and Δ' being on each side of it (Figure 4.6.1).

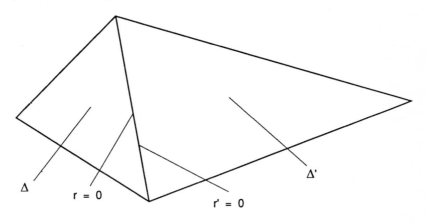

Figure 4.6.1

The parameters are then related by a relation of the following form

$$(4.6.1) \qquad \begin{cases} r' = -\alpha r \\ s' = s + \beta r \\ t' = t + \gamma r \end{cases} \qquad (\alpha > 0, \ 1 + \alpha = \beta + \gamma),$$

and to have a C^0 continuity, the curves image of $r = 0$ and $r' = 0$ have to coincide for B and B', which means that $P_{0,j,k} = P'_{0,j,k}$ $(j + k = n)$.

To have a C^1 continuity, the first derivatives of B and B' have to co-incide along the common curve Γ, the image under B or B' of the segment $r = r' = 0$. For fixed j, let g_j (resp. g'_j) be the affine map which sends Δ (resp. Δ') onto the triangle $T_j = \big(P_{0,j,k}, P_{0,j+1,k-1}, P_{1,j,k-1}\big)$ (resp. onto the triangle $T'_j = \big(P_{0,j,k}, P_{0,j+1,k-1}, P'_{1,j,k-1}\big)\big)$.

With the notation of 4.5.6, let us denote by $\overrightarrow{X}$ the vector $\overrightarrow{X_0 X_1}$. We then have, denoting by D_X the differentiation operator in the direction of the vector $\overrightarrow{X}$,

$$D_X B = n \sum_{i+j+k=n-1} B_{i,j,k}^{n-1}\big(\Delta_{r_0} P_{i+1,j,k} + \Delta_{s_0} P_{i,j+1,k} + \Delta_{t_0} P_{i,j,k+1}\big).$$

The barycentric coordinates of $\overrightarrow{X}$ are Δ_{r_0}, Δ_{s_0} and Δ_{t_0}, therefore we have

$$g_j(\overrightarrow{X}) = P_{0,j,k}\Delta_{r_0} + P_{0,j+1,k-1}\Delta_{s_0} + P_{1,j,k-1}\Delta_{t_0}.$$

Now, if x is a point of the segment $r = 0$ in Δ, it follows that

$$D_X B(x) = n \sum_{j+k=n-1} B_{0,j,k}^{n-1}(x) g_j(\overrightarrow{X}).$$

We have in the same way

$$D_X B'(x) = n \sum_{j+k=n-1} B_{0,j,k}^{n-1}(x) g_j'(\overrightarrow{X}).$$

The C^1 continuity condition along Γ between the patches B and B' is therefore the coincidence of the maps g_j and g_j' ($0 \le j \le n$), i.e., that $T_j \cup T_j'$ is the image of $\Delta \cup \Delta'$ under an unique affine map, which implies in particular that the triangles T_j and T_j' are *coplanar*.

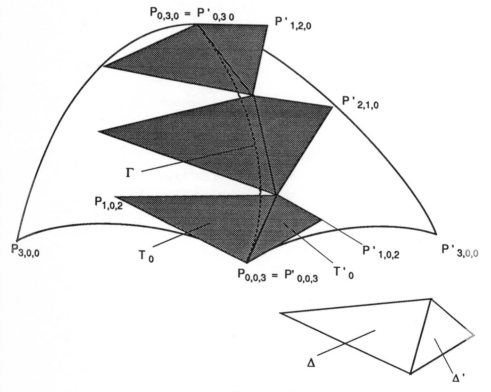

Figure 4.6.2

Figure 4.6.2 illustrates the C^1 continuity condition between two triangular patches of degree 3; the shaded triangles are coplanar.

4.6.2 Remarks

1) In the case when the patches are in $\mathbf{R}^3$ and projected on $\Delta \cup \Delta'$ by π, and when $\pi(P_{i,j,k}) \in \Delta$ has coordinates $(r = \frac{i}{n}, \ s = \frac{j}{n}, \ t = \frac{k}{n})$, (resp. $\pi(P'_{i,j,k}) \in \Delta'$ has coordinates $r' = \frac{i}{n}, \ s' = \frac{j}{n}, \ t' = \frac{k}{n}$), the $\mathcal{C}^1$ continuity condition is merely equivalent to the fact that the above triangles be coplanar.

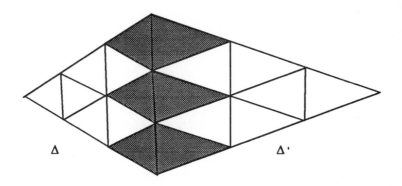

Figure 4.6.3

In Figure 4.6.3, to have a $\mathcal{C}^1$ continuity of the derivatives, the triangles whose projections are the shaded triangles have to be pairwise coplanar.

2) If we only want the tangent planes to the two patches to coincide along Γ (one says then that there is G_1 continuity), without necessarily $\mathcal{C}^1$ continuity of derivatives, the condition is less rigid, but less simple to express; see 4.2.5 above.

4.7 BASE POINTS OF RATIONAL BEZIER PATCHES

In this section, we will follow a paper of J. Warren ([W]), explaining the behavior of a triangular rational Bézier patch at a base point, and using this notion for the creation of multisided patches.

Let r, s, t be barycentric coordinates on a triangle $\Delta \subset \mathbf{R}^2$, and $B^n_{i,j,k}(r, s, t)$ the corresponding Bernstein basis (see 4.4.1). Let also P_{ijk} $(i + j + k = n)$ be control points in $\mathbf{R}^s$, and w_{ijk} $(i + j + k = n)$ be real numbers not all zero, called *weights*. As for the case of curves (see 3.5.6), one defines the rational triangular Bézier patches:

4.7.1 Definition   *The triangular rational Bézier patch associated to the points P_{ijk}'s and to the weights w_{ijk}'s is the image of the triangle Δ under the map*

$$R(r, s, t) = \frac{B(r, s, t)}{T(r, s, t)}$$

with

$$
\begin{cases}
B(r, s, t) = \displaystyle\sum_{i+j+k=n} P_{ijk} w_{ijk} B_{i,j,k}^n(r, s, t) \\[2mm]
T(r, s, t) = \displaystyle\sum_{i+j+k=n} w_{ijk} B_{i,j,k}^n(r, s, t).
\end{cases}
$$

For simplicity, we will assume that $w_{ijk} \geq 0$, which is the usual assumption in this domain.

4.7.2 Definition   *A base point of the rational Bézier patch $R(r, s, t)$ is a point (r_0, s_0, t_0) of Δ such that $B(r_0, s_0, t_0) = T(r_0, s_0, t_0) = 0$.*

4.7.3 Remarks

1) If $T(r_0, s_0, t_0) = 0$, but $B(r_0, s_0, t_0) \neq 0$, the image of the point (r_0, s_0, t_0) is defined if one allows points at infinity, i.e., if one assumes that the patch is in the projective space $\mathbf{RP}^s$.

2) The map $R(r, s, t)$ is not defined at a base point. In this section, we will look at the behavior of the patch near a base point; however, we will not treat the general case, but some special ones which may be useful for design, and which give a good insight into the general situation.

3) For example, if one has $w_{00n} = 0$, the point $(0, 0, 1)$ of Δ is a base point. In fact, one has then $B(0,0,1) = T(0,0,1) = 0$, since $B_{i,j,k}^n(r, s, t) = \frac{n!}{i!j!k!} r^i s^j t^k$, and therefore $B_{i,j,k}^n(0, 0, 1) = 0$ if $i > 0$ or $j > 0$.

Let us now define the blowing up of a point $Q \in \Delta$. For simplicity, we will assume that Q is one of the three vertices of Δ, say the point $(0, 0, 1)$. This is the most useful case for possible applications, as the blowing up of Q will let us interpret any triangular patch as a rectangular one (see below). For the general definition of blowing up, see [Fu] or [B-R].

4.7.4 Definition Let $Q = (0, 0, 1) \in \Delta$. The blowing up of Q is by definition the map ϕ of the square $C = [0, 1] \times [0, 1]$ (with cartesian coordinates (α, β)), onto Δ (with barycentric coordinates (r, s, t)) defined by

$$\phi \quad \begin{cases} r = \alpha(1 - \beta) \\ s = \alpha\beta \\ t = 1 - \alpha. \end{cases}$$

4.7.5 Remarks

1) A chart of the blowing up of the origin in $\mathbf{R}^2$ (with coordinates (x, y) is generally defined as the map: $\mathbf{R}^2 \longrightarrow \mathbf{R}^2$ defined by

$$\begin{cases} x = x' \\ y = x'y'. \end{cases}$$

This transformation is in fact the same as that of 4.7.4, if we make the change of variables

$$\begin{cases} r = x - y \\ s = y \\ t = 1 - x \end{cases} \quad \text{and} \quad \begin{cases} \alpha = x' \\ \beta = y'. \end{cases}$$

2) The map ϕ is invertible on $\Delta \setminus \{0\}$. In fact, the inverse ψ of ϕ is defined by the following formulas

$$\psi \quad \begin{cases} \alpha = r + s = 1 - t \\ \beta = \dfrac{s}{r + s}. \end{cases}$$

ψ is in fact defined on all the plane of Δ, except for the line $r + s = 0$, which intersects Δ only at the point Q.

Let us illustrate the map $\phi: \quad C \longrightarrow \Delta$.

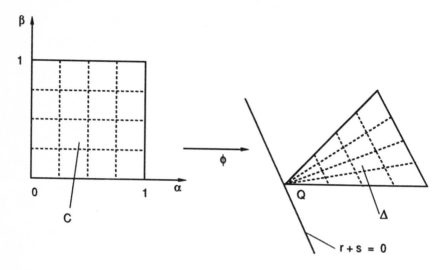

Figure 4.7.1

The image of the side $\alpha = 0$ of C is the point Q, and the images of the horizontal lines are the lines passing by Q. The map ϕ induces a bijection of $C \setminus \{\alpha = 0\}$ onto $\Delta \setminus \{Q\}$. This map is called the "blowing up" of Q, since the map ψ (defined on $\Delta \setminus \{Q\}$) "blows up" the point Q into the segment $\{\alpha = 0\}, \quad 0 \le \beta \le 1$ of C.

3) Let $R(r, s, t) : \Delta \longrightarrow \mathbf{R}^m$ be a rational Bézier patch of degree n. The map $R \circ \phi : C \longrightarrow \mathbf{R}^m$, which has the same image as ϕ, defines the patch as a rational Bézier tensor product patch of bidegree (n, n) defined over the unit square C. This patch has the same control polygon as R, but with modified weights (and with the image of a side degenerated to a point).

Let us illustrate this fact, as in [W], in the case $n = 2$. One has

$$B = w_{002}P_{002}t^2 + 2w_{101}P_{101}rt + w_{200}P_{200}r^2 + 2w_{110}P_{110}rs$$
$$+ w_{020}P_{020}s^2 + 2w_{011}st,$$
$$T = w_{002}t^2 + 2w_{101}rt + 2w_{011}st + w_{200}r^2 + 2w_{110}rs + w_{020}s^2$$

where $P_{ijk} \quad (i + j + k = 2)$ is the set of control points in $\mathbf{R}^m$.

Applying transformation ϕ yields the following tensor product rational quadratic surface

(4.7.1)

$$B \circ \phi(\alpha,\beta) = w_{002}P_{002}(1-\alpha)^2 + \big(w_{101}P_{101}(1-\beta)+w_{011}P_{011}\beta\big)\big(2\alpha(1-\alpha)\big)$$
$$+\big(w_{200}P_{200}(1-\beta)^2 + w_{110}P_{110}2(1-\beta)\beta + w_{020}P_{020}\beta^2\big)(\alpha)^2,$$

$$T \circ \phi(\alpha,\beta) = w_{002}(1-\alpha)^2 + \big(w_{101}(1-\beta)+w_{011}\beta\big)\big(2\alpha(1-\alpha)\big)$$
$$+\big(w_{200}(1-\beta)^2 + w_{110}2(1-\beta)\beta + w_{020}\beta^2\big)(\alpha)^2.$$

The constant terms, as in $w_{002}(1-\alpha)^2$, or the terms of degree one can be raised to quadratics to bring the patch into standard product form.

Let us now apply this transformation assuming Q is a base point (i.e., $w_{002}=0$ in this example). Then both $B \circ \phi$ and $T \circ \phi$ become divisible by α. Set

$$\begin{cases} \tilde{B}(\alpha,\beta) = \dfrac{B \circ \phi}{\alpha} \\ \tilde{T}(\alpha,\beta) = \dfrac{T \circ \phi}{\alpha} \end{cases}$$

Then the expression $\frac{\tilde{B}}{\tilde{T}}(\alpha,\beta)$ is equal to $\frac{B\circ\phi}{T\circ\phi}(\alpha,\beta)$ for $\alpha \neq 0$), and if w_{101} and w_{011} are non zero, it is defined for $\alpha = 0$. In the case of degree n, this defines a patch which is a tensor product of bidegree $(n-1,n)$ coinciding with the previous patch $\frac{B\circ\phi}{T\circ\phi}$ on $C \setminus \{\alpha = 0\}$, but which has four sides, the image of the segment $\alpha = 0$ being the line connecting $P_{1,0,n-1}$ and $P_{0,1,n-1}$: the triangular surface (with base point Q) is in fact four-sided.

More generally, if all weights w_{ijk} with $i+j < m$ are set to zero, $B \circ \phi$ and $T \circ \phi$ become divisible by α^m, and, after division of $B \circ \phi$ and $T \circ \phi$ by α^m, the triangular rational Bézier patch becomes four-sided, the fourth side (the image of $\alpha = 0$) being the rational Bézier curve of degree m with control points P_{ijk}, $i+j = m$, $i+j+k = n$, as is immediately seen from the expressions of $B \circ \phi$ and $T \circ \phi$ (see (4.7.1) for the case $n = 2$).

Let us see now how this property can be used to create multisided patches (for more details, the reader may refer to [W]). The technique described above (i.e., setting weights to zero to create a four-sided patch) may be generalized to create five- and six-sided patches, by treating each vertex of the triangular domain independently.

Figure 4.7.2 illustrates the creation of a six-sided patch, with degree $n = 4$; three portions of the boundary are formed with segments, and the others by rational quadratic Bézier curves.

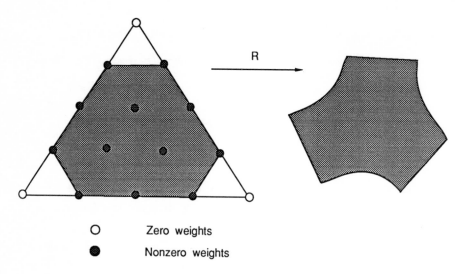

O   Zero weights

●   Nonzero weights

Figure 4.7.2

From a computational point of view, one has to treat the three vertices of Δ independently. A solution, proposed in [W], is to subdivide Δ in four triangles (and the patch in four patches), as shown in Figure 4.7.3.

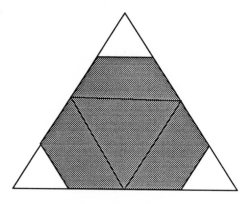

Figure 4.7.3

We will not study base points in the general case (for instance in the case where the weights are not assumed to be ≥ 0). It needs a little bit of the machinery of algebraic geometry (see [Fu]) to prove that the situation is similar to the one described above: after a sequence of "blowing up", the base points disappear, replaced by a union of lines in some projective space.

Let us now point out, as ever following [W], two interesting facts about base points. Let Δ be the triangle of parameters for a rational Bézier patch R defined by $R(r, s, t) = \frac{B(r,s,t)}{T(r,s,t)}$.

Definition   Let $M = \{(i/n, j/n, k/n)|w_{ijk} \neq 0\}$, and Π the convex hull of M. Π is contained in Δ, and is called the Newton polygon of R. The first result is a generalization of the previous method, allowing the creation of n-sided patches.

4.7.6 Theorem   *Let Π be the Newton polygon associated with the rational Bézier patch R. The boundary curves of R are in 1-1 correspondence with the edges of Π.*

Proof   Consider an edge e of $\partial\Pi$ joining the points $(i_1/n, j_1/n, k_1/n)$ and $(i_2/n, j_2/n, k_2/n)$. Assume that the line containing e separates Π and $Q = (0, 0, 1)$, and that $i_1 \geq i_2$ (and consequently $j_1 \leq j_2$). Then a similar transformation to ϕ, adapted to this situation, is

$$\tilde{\phi} \begin{cases} r = (1 - \beta)\alpha^{j_2 - j_1} \\ s = \beta\alpha^{i_1 - i_2} \\ t = 1 - r - s \end{cases}$$

(Similar transformations can be made for the points (1,0,0) and (0,1,0)).

$B \circ \tilde{\phi}$ and $T \circ \tilde{\phi}$ are now divisible by $\alpha^{i_1 j_2 - i_2 j_1}$, and after division by $\alpha^{i_1 j_2 - i_2 j_1}$, the image of the side $\{\alpha = 0\}$ is a side of the patch R, corresponding to the edge e. One has now to verify that the end points of two boundary curves of R corresponding to two consecutive edges of $\partial\Pi$ coincide, which is immediate considering the expression of $B \circ \tilde{\phi}$ and $T \circ \tilde{\phi}$.

Let us now look at the degree of the implicit equation of R. Note that it is very interesting in geometric design, for instance for all geometric problems like intersection, to obtain patches with an implicit equation of the lowest possible degree. We will see that the presence of base points may lower the degree of the implicit equation of the patch R.

Recall from Chapter 6 that R is a portion of an algebraic surface, i.e., that there exists a polynomial $Q(X, Y, Z) \in \mathbf{R}[X, Y, Z]$ of minimal degree such that the zero set of Q contains $R(\Delta)$.

4.7.7 Theorem   Let d be the degree of the implicit equation Q of the patch R. Then one has

$$d \leq \frac{\text{area}(\Pi)}{\text{area}(\Delta)} \, n^2.$$

The proof consists first in proving the following lemma

4.7.8 Lemma   With the above notation, let us assume that $\Pi = \Delta$. Then the degree of the implicit equation Q of the patch $R(\Delta)$ is $\leq n^2$.

Proof   The patch R is of the following form

$$\begin{cases} X = \dfrac{B_1(r,s,t)}{T(r,s,t)} \\[2mm] Y = \dfrac{B_2(r,s,t)}{T(r,s,t)} \\[2mm] Z = \dfrac{B_3(r,s,t)}{T(r,s,t)} \end{cases}$$

with the degree of the B_i's and of T being $\leq n$ by hypothesis. Let Z be the zero set of Q: one has $R(\Delta) \subset Z$, and it can be proved (see [Fu]) that, if we consider the extension of the situation over the field $\mathbf{C}$ of complex numbers, Z is the image of $\mathbf{C}^2$ (plus infinite points) by the map R. The degree of Q is then exactly the number of intersection points of Z with a generic line D, whose equations are of the form

$$D \quad \begin{cases} \alpha X + \beta Y + \gamma Z = c \\ \alpha' X + \beta' Y + \gamma' Z = c'. \end{cases}$$

For D and Z to intersect, we have to replace X, Y and Z by their values from the expression in $R(r,s,t)$; we obtain therefore two homogeneous equations of degree $\leq n$ in (r,s,t), which define two algebraic curves in the projective plane $\mathbf{RP^2}$ of degree $\leq n$, and which have at most n^2 intersection points, by Bezout's theorem (see [Fu], or [B-R]).

∎

The complete proof of the theorem is analoguous in the homogeneous case to the Bernstein-Kushnirenko theorem (see [Be]) which relates the number of solutions (in $\mathbf{C_*}^n$) of an algebraic system of n equations in n unknowns to the volume of their Newton polyhedra.

Let us give an example of an application of this theorem. Take $n = 6$, and set $w_{ijk} = 0$ if $i > 3$ or $j > 3$. Then the Newton polygon Π is shown in

Figure 4.7.4, and we have $\frac{\text{area}(\Pi)}{\text{area}(T)} = \frac{1}{2}$.

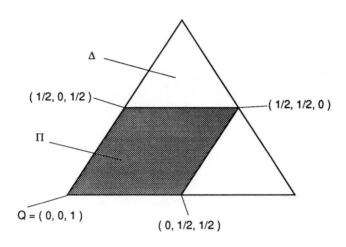

Figure 4.7.4

Let us now take a rational bicubic Bézier patch $S(X, Y)$. The transformation

$$\begin{cases} X = \dfrac{r}{t} \\ Y = \dfrac{s}{t} \end{cases}$$

applied to $S(X, Y)$ gives a triangular patch $R(r, s, t)$ of degree six, with Newton polygon contained in Π. These two patches have the same implicit equation (since they have the same image in the projective space $\mathbf{RP}^3$).

One sees therefore, applying Theorem 4.7.7, that any bicubic rational polynomial patch (for instance in Bézier form), has an implicit equation of degree ≤ 18 (and not 36, which would be expected from Lemma 4.7.8).

4.8 POLYHEDRAL SPLINES

We will define polyhedral splines and study their fundamental properties; the reader may refer to [D-M] for a relatively complete study of the subject.

The formula proved in 1.7.6 suggests the following generalization of the B-splines to the s-variable case.

Let $X_0, \ldots, X_n$ be $n+1$ points in $\mathbf{R}^s$, and let $[X_0 \ldots X_n]$ be their convex hull; we will assume $\mathrm{Vol}[X_0 \ldots X_n] > 0$, which implies $n \geq s$ (Vol means the s-dimensional volume in $\mathbf{R}^s$).

4.8.1 Definition   *Let $S_n \subset \mathbf{R}^{n+1}$ the standard simplex of dimension n. We   define   the   B-spline   associated   to $X_0, \ldots, X_n$,   denoted $B(x|X_0, \ldots, X_n)$, as the unique function on $\mathbf{R}^s$ satisfying*

$$\int_{\mathbf{R}^s} f(x) B(x|X_0, \ldots, X_n) = n! \int_{S_n} f(t_0 X_0 + \cdots + t_n X_n) \, dt_1 \ldots dt_n$$

for any function f of class C^∞ with compact support on $\mathbf{R}^s$.

4.8.2 Remarks

a) Definition 4.8.1 introduces B as a *distribution*. We will see below that, with the hypotheses we have made, B is in fact a function on $\mathbf{R}^s$ (see for instance 4.8.7 below).

b) As in Chapter 1, Section 7, the notation

$$\int_{S_n} f(t_0 X_0 + \cdots + t_n X_n) \, dt_1 \ldots dt_n$$

means that we set $t_0 = 1 - t_1 - \cdots - t_n$, and that we integrate on the projection S'_n of S_n in $\mathbf{R}^n$: see Figure 4.8.1. $S'_n \subset \mathbf{R}^n$ is defined by $t_i \geq 0$, $\sum_{i=1}^n t_i \leq 1$; this fixes in particular the orientation of S_n. We may moreover replace t_0 by any t_j ($0 \leq j \leq n$), multiplying the result by $(-1)^j$, in such a way that the orientation is preserved.

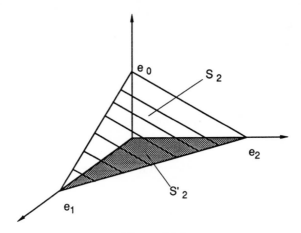

Figure 4.8.1

c) Let $\pi : \mathbf{R}^n \longrightarrow \mathbf{R}^s$ be the affine map defined by

$$\pi(e_i) = X_i - X_0 \quad (1 \leq i \leq n), \quad \pi(0) = X_0,$$

$e_1, \ldots, e_n$ denoting the canonical basis of $\mathbf{R}^n$. Then, if $\tilde{\pi}$ denotes the restriction of π to S'_n, the preceding definition introduces $B(x|X_0, \ldots, X_n)$ (viewed as a distribution) as the direct image $\tilde{\pi}_* \omega$, where ω is the differential form $\omega = n! \, dt_1 \wedge \ldots \wedge dt_n$. This form satisfies to the relation $\int_{S'_n} \omega = 1$, which justifies the factor $n!$.

The preceding definition suggests a generalization to the case where the simplex S_n (or S'_n) is replaced by any polyhedron.

4.8.3 Definition *Let $P \subset \mathbf{R}^n$ be a compact polyhedron, $\pi : \mathbf{R}^n \longrightarrow \mathbf{R}^s$ an affine map. One defines $B_P(x)$, the polyhedral spline associated to P (and to π), by the formula*

$$\int_{\mathbf{R}^s} f(x) B_P(x) = \int_P f \circ \pi \, dt_1 \wedge \ldots \wedge dt_n$$

for any function f of class $\mathcal{C}^\infty$ with compact support on $\mathbf{R}^s$.

Note that the B-spline $B_P(x)$ is defined by the above formula only as a distribution; we will see below under which conditions $B_P(x)$ is in fact a function. The B-spline defined in 4.8.1 then corresponds, up to the factor

$n!$, to the case where P is the projection S'_n of S_n in $\mathbf{R}^n$. We thus have $B(x|X_0,\ldots,X_n) = n! \, B_{S'_n}(x)$. The factor $n!$ introduced here is classical: see for instance [D-M].

The preceding definition implies easily

4.8.4 Proposition

a) *The support of* $B_P(x)$ *is equal to* $\pi(P)$ *(in particular, the support of* $B(x|X_0,\ldots,X_n)$ *is the convex hull* $[X_0\ldots X_n]$ *of the* X_i*'s).*

b) *If* π *is surjective, we have* $B_P(x) = \mathrm{Vol}_{n-s}(\pi^{-1}(x)\cap P)$; *this proves in particular that* $B_P(x)$ *is a function if* $\mathrm{Vol}_{n-s}(\pi^{-1}(x)\cap P)$ *is defined, i.e., as soon as* $\pi(P)$ *generates* $\mathbf{R}^s$, *as is so in Definition 4.8.1.*

c) *In the case where* $n = s$,

$$B(x|X_0,\ldots,X_n) = \frac{\chi_{[X_0\ldots X_n]}(x)}{\mathrm{Vol}([X_0\ldots X_n])},$$

χ *denoting the characteristic function.*

Proof

a) The support of $B_P(x)$ is by definition the complement of the set of points $x \in \mathbf{R}^s$ in the neighborhood of which $B_P(x)$ is identically zero. In other words (considering B_P as a distribution), x is in the complement U of the support of B_P if and only if B_P is zero on any function with support in U: it is clear that the complement of $\pi(P)$ satisfies this condition.

b) Assume π surjective. We may then take coordinates $(t_1,\ldots t_n)$ on $\mathbf{R}^n$ and $(t_1\ldots t_s)$ on $\mathbf{R}^s$ such that $\pi(t_1,\ldots,t_n) = (t_1,\ldots,t_s)$. We have then

$$\int_P f\circ\pi \, dt_1 \wedge\ldots\wedge dt_n = \int_p f(t_1,\ldots,t_s) dt_1 \wedge\ldots\wedge dt_n$$

$$= \int f(t_1,\ldots t_s)\left[\int_{\pi^{-1}(x)\cap P} dt_{s+1}\wedge\ldots\wedge dt_n\right] dt_1\wedge\ldots\wedge dt_s$$

denoting by x the point with coordinates $(t_1,\ldots,t_s)$.

If $\pi^{-1}(x)$ is of dimension $n-s$, we have

$$\int_{\pi^{-1}(x)\cap P} dt_{s+1}\wedge\ldots\wedge dt_n = \mathrm{vol}_{n-s}(\pi^{-1}(x)\cap P)$$

which proves that in this case, $B_P(x) = \mathrm{Vol}_{n-s}(\pi^{-1}(x)\cap P)$.

c) Assume $n = s$ and $\mathrm{Vol}[X_0, \ldots, X_n] > 0$. π is then a linear bijection of $\mathbf{R}^n$, i.e., a linear change of variables. The formula of change of variables (for the map π^{-1}) gives then

$$\int_{S'_n} f \circ \pi \, dt_1 \wedge \ldots \wedge dt_n = \int_{\mathbf{R}^s} (f \circ \pi) \chi(S'_n) dt_1 \wedge \ldots \wedge dt_n$$

$$= \frac{1}{\mathrm{Vol}[X_0, \ldots, X_n]} \int_{\mathbf{R}^n} f \chi([X_0, \ldots, X_n]),$$

which completes the proof of the proposition.

∎

Let us now prove two fundamental properties of the polyhedral splines: the differentiation formula, and the induction formula. For simplicity, we will restrict ourselves to the case of B-splines (a polyhedral spline is in fact always a sum of B-splines, obtained by a triangulation of P, i.e., by a decomposition of P into a union of simplices; note however that this decomposition is not canonical, the triangulation of P not being unique).

Let us denote by $D_{X_i - X_j}$ the derivative in the direction of the vector $X_i - X_j$, and by $(X_0, \ldots, \widehat{X_j}, \ldots, X_n)$ the sequence $(X_0, \ldots, X_n)$ in which the point X_j has been omitted. With this notation, we have

4.8.5 Proposition

$$D_{X_i - X_j} B(x|X_0, \ldots, X_n)$$
$$= n \Big(B(x|X_0, \ldots, \widehat{X_i}, \ldots, X_n) - B(x|X_0, \ldots, \widehat{X_j}, \ldots X_n) \Big).$$

Proof By linearity, we can assume that $j = 0$ (as we have $D_{X_i - X_j} = D_{X_i - X_0} + D_{X_0 - X_j}$). We have

$$\int_{\mathbf{R}^s} D_{X_i - X_0} B(x|X_0, \ldots, X_n) f(x) = -\int_{\mathbf{R}^s} B(x|X_0, \ldots, X_n) D_{X_i - X_0} f(x)$$

by "integration by parts" (f having a compact support), or alternatively by definition of B considered as a distribution, and using

$$\int_{\mathbf{R}^s} B(x|X_0, \ldots, X_n) D_{X_i - X_0} f(x) = n! \int_{S'_n} D_{e_i}(f \circ \pi) \, dt_1 \wedge \ldots \wedge dt_n,$$

by definition of $B(x|X_0, \ldots, X_n)$, using the fact that π is affine. We have then

$$\int_{S'_n} D_{e_i}(f \circ \pi) \, dt_1 \wedge \ldots \wedge dt_n = \int_{S'_n} \frac{\partial}{\partial t_i}(f \circ \pi) \, dt_1 \wedge \ldots \wedge dt_n$$

S'_n being, as above, the projection of S_n in $\mathbf{R}^n$. Set

$$\omega = (-1)^{i-1} n! (f \circ \pi)\, dt_1 \wedge \ldots \wedge \widehat{dt_i} \wedge \ldots \wedge dt_n.$$

We have then

$$d\omega = n! \frac{\partial}{\partial t_i} (f \circ \pi)\, dt_1 \wedge \ldots \wedge dt_n,$$

and Stokes's formula (immediate in the case of a simplex) is written

$$\int_{S'_n} d\omega = \int_{\partial S'_n} \omega$$

where $\partial S'_n$ denotes the boundary of the simplex S'_n (see Figure 4.8.2 for the case $n = 2$).

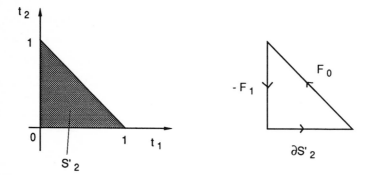

Figure 4.8.2

There are only two faces of S'_n on which the differential form ω is not identically zero: the face F_0 defined by $\sum_{i=1}^{n} t_i = 1$, $t_i \geq 0$ and the face F_i defined by $t_i = 0$. We have therefore

$$\int_{S'_n} d\omega = \int_{\partial S'_n} \omega = \int_{F_0} \omega + \int_{F_i} \omega.$$

But the face F_0, is equal to S_{n-1}, with the same orientation. We have therefore

$$\int_{F_0} \omega = (-1)^{i-1} n! \int_{S_{n-1}} (f \circ \pi) dt_1 \wedge \ldots \wedge \widehat{dt_i} \wedge \ldots \wedge dt_n$$

$$= (-1)^{i-1} \times (-1)^{i-1} n \int_{\mathbf{R}^s} f(x) B(x | X_1, \ldots, X_n);$$

the second equality results from the definition of $B(x|X_1,\ldots,X_n)$: see Remark 4.8.2, b). We find in the same way

$$\int_{F_i} \omega = (-1)^{i-1} n! \int_{F_i} (f \circ \pi) dt_1 \wedge \ldots \wedge \widehat{dt_i} \wedge \ldots \wedge dt_n$$

$$= (-1)^{i-1} \times (-1)^i n \int_{\mathbf{R}^s} f(x) B(x|X_0,\ldots,\widehat{X_i},\ldots,X_n).$$

The sign $(-1)^i$ comes from the fact that the face F_i is oriented in the same way as the boundary of S'_n, and this orientation is equal to $(-1)^i$ times that of the hyperplane $\mathbf{R}^{n-1}$ defined by $X_i = 0$ (see Figure 4.7.2). We have then finally

$$\int_{\mathbf{R}^s} f(x) D_{X_i - X_0} B(x|X_0,\ldots,X_n)$$

$$= -\int_{S'_n} d\omega = -\int_{\partial S'_n} \omega = -\left(\int_{F_0} \omega + \int_{F_i} \omega\right)$$

$$= n\left(\int_{\mathbf{R}^s} f(x) B(x|X_0,\ldots,\widehat{X_i},\ldots,X_n) - \int_{\mathbf{R}^s} f(x) B(x|X_1,\ldots,X_n)\right).$$

∎

4.8.6 Corollary *The function $B(x|X_0,\ldots,X_n)$ is piecewise polynomial of degree $\le n-s$, the pieces being defined by the partition of $[X_0 \ldots X_n]$ generated by the mutual intersections of sets of the form $[X_{i_0} \ldots X_{i_s}]$ $(0 \le i_j < i_{j+1} \le n)$.*

Proof We use induction on n, the case $n = s$ resulting from the fact that

$$B(x|X_0,\ldots,X_s) = \frac{\chi_{[X_0\ldots X_s]}(x)}{\mathrm{Vol}([X_0 \ldots X_s])},$$

and the step from n to $n+1$ being an immediate consequence of Proposition 4.8.5.

∎

4.8.7 Corollary *$B(x|X_0,\ldots,X_n)$ is a function on $\mathbf{R}^s$ of class C^{d-2}, if d is the smallest of the cardinalities of the subsets $Y \subset X = (X_0,\ldots,X_n)$ such that $X \setminus Y$ does not affinely span $\mathbf{R}^s$.*

Proof Corollary 4.8.7 is equivalent to the following property: "if all the subsets of p points (among $(X_0, \ldots, X_n)$) generate $\mathbf{R}^s$, then $B(x|X_0, \ldots, X_n)$ is of class $\mathcal{C}^{n-p}$".

In fact, let Z be a set of p points generating $\mathbf{R}^s$, with p minimal. Then $X \setminus Z$ has $n + 1 - p$ elements, and removing a point from Z (i.e., adding a point to $X \setminus Z$), we obtain a part Y with $d = n + 2 - p$ elements. Corollary 4.8.7 is now an easy consequence of Proposition 4.8.5, because to define $B(x|X_0, \ldots, X_n)$, we have assumed $\mathrm{Vol}[X_0, \ldots, X_n] > 0$, i.e., that $X = (X_0, \ldots, X_n)$ generates $\mathbf{R}^s$.

■

4.8.8 Corollary   Let $Z = \sum_{i=0}^n \mu_i X_i$, with $\sum_{i=0}^n \mu_i = 0$. Then

$$D_Z B(x|X_0, \ldots, X_n) = n \sum_{j=0}^n \mu_j B(x|X_0, \ldots, \widehat{X_j}, \ldots, X_n).$$

Proof Immediate by 4.8.5.

■

4.8.9 Remarks

a) In the case $s = 1$, we recognize the properties of the one variable B-spline functions of degree $n - 1$.

b) We see that when the X_i's are in *general position*, $B(x|X_0, \ldots, X_n)$ is of class $\mathcal{C}^{n-s-1}$ (general position meaning here that all the subsets of $(X_0, \ldots, X_n)$ with $s + 1$ elements generate $\mathbf{R}^s$).

Let us now prove the induction relation on the B-splines, which generalizes to polyhedral splines the relation 1.3.2 concerning the one-variable B-splines.

4.8.10   Proposition   Let x be   a   point   in $[X_0 \ldots X_n]$, $x = \sum_{j=0}^n \lambda_j X_j$, $\sum_{j=0}^n \lambda_j = 1$. Then

$$B(x|X_0, \ldots, X_n) = \frac{n}{n - s} \sum_{j=0}^n \lambda_j B(x|X_0, \ldots, \widehat{X_j}, \ldots, X_n).$$

Proof Note first that the constant $\frac{n}{n-s}$ comes from the fact that Definition 4.8.1 is a generalization of the one-variable B-splines multiplied by a constant (compare 4.8.1 and 1.7.6).

Note next that the λ_j's are in general not uniquely determined.

Let (e_i) $(1 \leq i \leq n)$ be the canonical basis of $\mathbf{R}^n$, and $\pi : \mathbf{R}^n \longrightarrow \mathbf{R}^s$ the affine map defined by $\pi(e_i) = X_i - X_0$ $(1 \leq i \leq n)$; set $Y = \sum_{i=1}^{n} \lambda_i e_i$ (we have then $Y \in \pi^{-1}(x)$). The $\lambda_i's$ $(0 \leq i \leq n)$ are then the barycentric coordinates of Y with respect to the points $O, e_1, \ldots, e_n$, and $\lambda_i/n!$ is the volume of the simplex $\tilde{S}_i = [YO\, e_1 \ldots \hat{e}_i \ldots e_n]$ (see Figure 4.8.3).

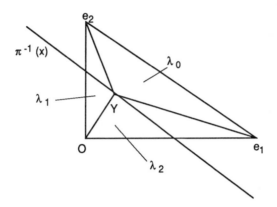

Figure 4.8.3

Proposition 4.8.4, b) then implies immediately

$$B(x|X_0, \ldots, X_n) = n!\mathrm{Vol}\big(\pi^{-1}(x) \cap S_n'\big) = n! \sum_{i=0}^{n} \mathrm{Vol}\big(\pi^{-1}(x) \cap \tilde{S}_i\big)$$

$$= \sum_{i=0}^{n} \lambda_i B(x|x, X_0, \ldots, \widehat{X}_i, \ldots, X_n)$$

since, if ϕ denotes the canonical linear map: $S_n' \longrightarrow \tilde{S}_i$, we have

$$n!\mathrm{Vol}\big(\pi^{-1}(x) \cap \tilde{S}_i\big) = n! \times \frac{\mathrm{Vol}\tilde{S}_i}{\mathrm{Vol}S_n'} \mathrm{Vol}\big((\pi \circ \phi)^{-1}(x)\big)$$

$$= n!\lambda_i \mathrm{Vol}\Big((\pi \circ \phi)^{-1}(x)\Big) = \lambda_i B(x|x, X_0, \ldots, \widehat{X}_i, \ldots, X_n).$$

We have now only to prove that

$$B(x|x, X_0, \ldots, \widehat{X}_i, \ldots, X_n) = \frac{n}{n-s} B(x|X_0, \ldots, \widehat{X}_i, \ldots, X_n),$$

which, after a change of notation, is the following lemma.

4.8.11 Lemma

$$B(X_0|X_0,\ldots,X_n) = \frac{n}{n-s}B(X_0|X_1,\ldots,X_n).$$

Proof Let us come back to the definition of $B(X_0|X_0,\ldots,X_n)$ as the volume of $(\pi^{-1}(X_0) \cap S'_n)$ multiplied by $n!$ (see 4.8.4, b). S'_n is a cone on S_{n-1} of vertex O, and we have

$$\frac{B(X_0|X_0,\ldots,X_n)}{B(X_0|X_1,\ldots,X_n)} = \frac{n!\mathrm{Vol}(\pi^{-1}(X_0)\cap S'_n)}{(n-1)!\,\mathrm{Vol}(\pi^{-1}(X_0)\cap S_{n-1})}$$

$$= n\int_0^1 (1-t)^{n-s-1}dt = \frac{n}{n-s}$$

(see Figure 4.8.4).

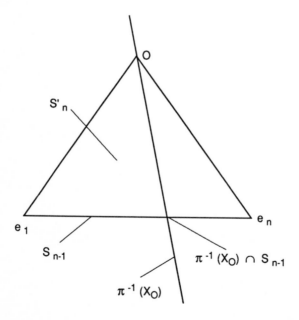

Figure 4.8.4

Application: Bernstein polynomials

Let $X_0, \ldots, X_s$ in $\mathbf{R}^s$ be given, each X_i being of "multiplicity $m_i + 1$" (i.e., counted $m_i + 1$ times), with $\sum_{i=0}^{s}(m_i + 1) = n + 1$. Then, if λ_i $(0 \leq i \leq s)$ are such that

$$X = \sum_{i=0}^{s} \lambda_i X_i, \quad \sum \lambda_i = 1,$$

we have, with obvious notation

4.8.12 Corollary

$$B(x|(X_0)^{m_0+1}, \ldots, (X_s)^{m_s+1}) = n! \frac{\lambda_0^{m_0}}{m_0!} \cdots \frac{\lambda_s^{m_s}}{m_s!} \frac{\chi_{[X_0 \ldots X_s]}(x)}{\text{Vol}[X_0 \ldots X_s]}.$$

In other words, we recognize, up to a constant factor, the generalization to dimension s of Bernstein polynomials of degree $\sum_{i=0}^{s} m_i = n - s$ defined in 4.4.1. Note that the λ_i's are the barycentric coordinates of x with respect to the X_i's, and so are here uniquely determined.

Proof We have only to apply $n-s$ times the formula from Proposition 4.7.10, "deleting" m_i times each X_i $(\sum_{i=0}^{s} m_i = n - s)$. We get then, by 4.8.4, c):

$$B\big(x|(X_0)^{m_0+1}, \ldots, (X_s)^{m_s+1}\big)$$
$$= \frac{n(n-1)\ldots(s+1)}{(n-s)!} \lambda_0^{m_0} \cdots \lambda_s^{m_s} \frac{\chi_{[X_0 \ldots X_s]}}{\text{Vol}[X_0 \ldots X_s]} N,$$

where N is the number of ways to pass from the sequence $(m_0+1, \ldots, m_s+1)$ to the sequence $(1, \ldots, 1)$, namely

$$\frac{(m_0 + \cdots + m_s)!}{m_0! \ldots m_s!} = \frac{(n-s)!}{m_0! \ldots m_s!}.$$

(N is the coefficient of the monomial $X_0^{m_0} \ldots X_s^{m_s}$ in the expansion of $(X_0 + \cdots + X_s)^{\sum m_i}$).

■

4.9 BOX SPLINES

We will consider a particular case of polyhedral splines, when the polyhedra P is the cube $[0,1[^n \subset \mathbf{R}^n$. We may then look at this spline as a generalization of the one variable spline $B_n(x)$ (defined on the interval $[0, n+1]$: see 2.3, f)), and in particular consider the space generated by the translates (by vectors in the lattice $\mathbf{Z}^n$) of the hypercube P. These splines are then called Box-splines.

Let X be a given set of $n + 1$ points $X_0, X_1, \ldots, X_n$ in $\mathbf{R}^s$, satisfying the following assumptions:

a) $X_0 = 0$, and the X_i's have integer coordinates, several X_i's being possibly equal.

b) The X_i's generate $\mathbf{R}^s$ (in this section, we will identify the points X_i $(i \geq 1)$ with the vectors $X_i - X_0$ of $\mathbf{R}^s$).

4.9.1 Definition   *Let* $(e_1, \ldots, e_n)$ *be the canonical basis of* $\mathbf{R}^n$, $\pi : \mathbf{R}^n \longrightarrow \mathbf{R}^s$ *the linear map defined by* $\pi(e_i) = X_i$ $(1 \leq i \leq n)$, $P = [0,1[^n \subset \mathbf{R}^n$. *Then the spline* $B_P(x)$ *associated with these data (see 4.8.3) is called the box-spline associated to* X *and is denoted* $M(x|X)$.

To mark a difference with the B-splines of Section 4.8, we have avoided the classical notation $B(x|X)$

We have then by definition

$$\int_{\mathbf{R}^s} f(x) M(x|X) = \int_{[0,1[^n} (f \circ \pi) dt_1 \wedge \ldots \wedge dt_n$$

for all f that are C^∞ with compact support.

4.9.2 Example $n = 3, s = 2$.

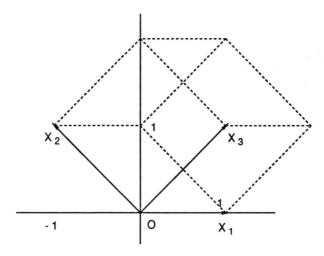

Figure 4.9.1

The edges of the projection of P are the vectors X_1, X_2 and X_3, and the edges in dotted lines; $\pi(P)$ is the support of $B_P(x) = M(x|X)$.

4.9.3 Remarks

a) Let us denote by X the set of X_i's $(1 \leq i \leq n)$; if X does not generate $\mathbf{R}^s$, $M(x|X)$ is, as in Section 4.8, a distribution on $\mathbf{R}^s$, with support $[X_0, \ldots, X_n]$. However, when we write the expression $M(x|X)$, we will always assume that either X generates $\mathbf{R}^s$, or $x \notin [X_0 \ldots X_n]$, in which case we will set $M(x|X) = 0$.

b) In the case where X does not generate $\mathbf{R}^s$, the distribution $M(x|X)$ on $\mathbf{R}^s$ may also be interpreted as a box-spline function on the linear subspace H of $\mathbf{R}^s$ generated by X, setting

$$\int_H f(x)M(x|X) = \int_{[0,1[^n} (f \circ \pi)dt_1 \wedge \ldots \wedge dt_n$$

for any function f on H of class C^∞ with compact support.

Let us now look at the properties of $M(x|X)$ similar to 4.8.5 and 4.8.10 (Proposition 4.8.4 remains without any change).

Let f be a function on $\mathbf{R}^s$, and set

$$\begin{cases} \Delta_{X_i} f = f(x + X_i) - f(x) \\ \nabla_{X_i} f = f(x) - f(x - X_i). \end{cases}$$

We have then

4.9.4 Proposition   *If for $i > 0$ D_{X_i} denotes the differentiation operator in the direction of the vector X_i, and if $X \setminus X_i$ generates $\mathbf{R}^s$, $D_{X_i}M(x|X)$ exists, and we have:*

$$D_{X_i}M(x|X) = \nabla_{X_i}M(x|X \setminus X_i).$$

Proof   As in 4.8.5, we have

$$\int_{\mathbf{R}^s} D_{X_i}M(x|X)f(x) = -\int_{\mathbf{R}^s} M(x|X)D_{X_i}f(x)$$

$$= -\int_{[0,1[^n} D_{X_i}(f \circ \pi)dt_1 \wedge \ldots \wedge dt_n$$

$$= (-1)^i \int_{\partial[0,1[^n} (f \circ \pi)dt_1 \wedge \ldots \wedge \widehat{dt_i} \wedge \ldots \wedge dt_n,$$

the last equality being the Stokes formula: $\int_P d\omega = \int_{\partial P} \omega$, with $P = [0,1[^n$, and $\omega = (f \circ \pi)dt_1 \wedge \ldots \wedge \widehat{dt_i} \wedge \ldots \wedge dt_n$. But the form ω is nonzero only on the faces F_i (defined by $t_i = 0$) and F_i' (defined by $t_i = 1$), since on each of the other faces, one of the t_j's ($j \neq i$) is constant. Because F_i and F_i' are oriented in the same way as the boundary of $[0,1[^n$ (see Figure 4.9.2), and then have opposite orientations (the orientation of F_i as the boundary of P is $(-1)^i$ times that of the hyperplane $\mathbf{R}^{n-1}$ with the basis $(e_1, \ldots, \widehat{e_i}, \ldots, e_n)$), we therefore find finally

$$D_{X_i}M(x|X) = M(x|X \setminus X_i) - M(x - X_i|X \setminus X_i) = \nabla_{X_i}M(x|X \setminus X_i).$$

■

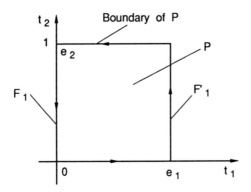

Figure 4.9.2 : n = 2

4.9.5 Corollary   *The function $M(x|X)$ is a piecewise polynomial of degree $\leq n - s$, and of class C^{d-1}, if $d + 1$ is the smallest of the cardinalities of the subsets $Y \subset X = (X_1, \ldots, X_n)$ such that $X \setminus Y$ does not generate $\mathbf{R}^s$ (as a vector space). (In other words, for any subset $Y \subset X$ with cardinality $\leq d$, $X \setminus Y$ generates $\mathbf{R}^s$.)*

   Proof   Immediate from Proposition 4.9.4.

                                                                           ∎

   We have also, as in 4.8.10, an induction relation

4.9.6 Proposition   *Let x be a vector of $\mathbf{R}^s$; we may then write, in a non-unique way if $n > s$, $x = \sum_{i=1}^{n} t_i(X_i - X_0)$. We have then (using Remark 4.9.3)*

$$M(x|X) = \frac{1}{n-s} \sum_{i=1}^{n} \left[t_i M(x|X \setminus X_i) + (1 - t_i) M(x - X_i|X \setminus X_i) \right].$$

   Proof   The proof is similar to that of Proposition 4.8.10, with the simplex S'_n replaced by the hypercube $[0, 1[^n$ (Figure 4.9.3). We take $t \in \mathbf{R}^n$, $t = \sum_{i=1}^{n} t_i e_i$, and we have

$$M(x|X) = \mathrm{Vol}\left(\pi^{-1}(x) \cap [0, 1[^n \right)$$

$$= \sum_{i=1}^{n} \mathrm{Vol}\left(\pi^{-1}(x) \cap C_i \right) + \sum_{i=1}^{n} \mathrm{Vol}\left(\pi^{-1}(x) \cap C'_i \right),$$

C_i (resp. C'_i) being the cone with vertex t and base equal to the face F_i (resp. F'_i).

■

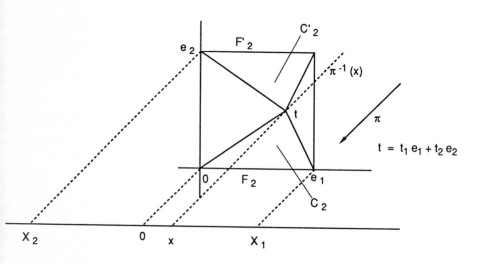

Figure 4.9.3 : $n = 2$, $s = 1$

We will now consider the vector space S_X generated by the functions $M(x - \alpha|X)$, where $\alpha \in \mathbf{Z}^s$ with the following meaning: S_X is the set of functions

$$\sum_{\alpha \in \mathbf{Z}^s} \lambda_\alpha M(x - \alpha|X), \quad \lambda_\alpha \in \mathbf{R}$$

(we thus allow infinite sums, which are meaningful, as the family of the $M(x|X)$'s is locally finite). This space is a possible generalization to the case of s variables, of the space generated by the (one variable) uniform B-splines with integer knots.

Two questions arise in a natural way:

a) If $\mathbf{P}_r$ denotes the space of polynomials of degree $\le r$ on $\mathbf{R}^s$, determine the integers r such that $\mathbf{P}_r \subset S_X$.

b) Study the linear independence of the functions $M(x - \alpha|X)$.

Let us look first at question a). It is important, since it determines the possible order of approximation of the functions on $\mathbf{R}^s$ by elements of S_X.

4.9.7 Proposition *Let d be the integer defined in 4.9.5; we then have* $\mathbf{P}_d \subset S_X$.

Remark In the "generic" case, i.e., in the case where any part of X with s elements generates $\mathbf{R}^s$, we have $d = n - s$, as for the one variable splines. For instance, in the case treated in 4.9.2, we have $d = 1$.

Proof If Y is a subset of $X = (X_1, \ldots, X_n)$, consider the differential operator $D_Y = \prod_{y \in Y} D_y$, where D_y is the differentiation operator in the direction of the vector y. The proof of Proposition 4.9.7 uses the following lemma.

4.9.8 Lemma *Let:*

$$Y(X) = \{Y \subset X | X \setminus Y \quad \text{does not generate} \quad \mathbf{R}^s\},$$

$$\mathcal{D}(X) = \{f \in C^\infty(\mathbf{R}^s) | D_Y f = 0 \ \forall \ Y \in Y(X)\}.$$

We then have $\mathcal{D}(X) \subset S_X$.

Let us complete the proof of the proposition, admitting Lemma 4.9.8 for the moment. Let Q be a polynomial of degree less than or equal to d. Then $Q \in \mathcal{D}(X)$ by definition of $\mathcal{D}(X)$, since if $Y \in Y(X)$, $\operatorname{card} Y \geq d + 1$, and then $D_Y Q = 0$. We then have $Q \in S_X$ by Lemma 4.9.8. ∎

Proof of Lemma 4.9.8

We will argue by induction on $n = \operatorname{card} X$; let $X_i \in X$.

α) Assume that $X \setminus X_i$ generates $\mathbf{R}^s$ and set $X' = X \setminus X_i$. Then, for any $y' \in Y(X')$, we have $D_{Y'} D_{X_i} f = 0$, because $Y' \cup \{X_i\} \in Y(X)$, since by hypothesis, $X' \setminus Y'$ does not generate $\mathbf{R}^s$ (4.9.6), and $X' \setminus Y' = X \setminus (Y' \cup \{X_i\})$. The induction hypothesis implies that $D_{X_i} f \in S_{X'}$, and we may write

$$D_{X_i} f = \sum_{\alpha \in \mathbf{Z}^s} a_\alpha M(x - \alpha | X \setminus X_i).$$

But there exist coefficients c_α ($\alpha \in \mathbf{Z}^s$), non unique in general, such that we can write

$$a_\alpha = c_\alpha - c_{(\alpha - X_i)} \quad \forall \alpha \in \mathbf{Z}^s.$$

(We have only to argue by induction on s, beginning with $s = 1$, and setting for instance $c_0 = 0$, which determines the coefficients c_α, $\alpha \in \mathbf{Z}$).

We then have

$$D_{X_i}f = \sum_{\alpha \in \mathbf{Z}^s} (c_\alpha - c_{(\alpha - X_i)})M(x - \alpha|X \setminus X_i)$$

$$= \sum_{\alpha \in \mathbf{Z}^s} c_\alpha \big(M(x - \alpha|X \setminus X_i) - M(x - \alpha - X_i|X \setminus X_i)\big)$$

$$= \sum_{\alpha \in \mathbf{Z}^s} c_\alpha D_{X_i}M(x - \alpha|X).$$

We may therefore write $f = \phi + \psi$, with $\phi \in S_X$, and $D_{X_i}\psi = 0$.

β) If $X \setminus X_i$ does not generate $\mathbf{R}^s$, we have $D_{X_i}f = 0$ by hypothesis.

We may apply the same argument again with the function ψ and another vector X_j of X, until we obtain a constant function, as by hypothesis, X generates $\mathbf{R}^s$; it is then enough to prove that $1 \in S_X$, which is the particular case $d = 0$ of 4.9.7. But $\sum_\alpha M(x - \alpha|X)$ is a constant independent of x, equal to the volume of $\pi^{-1}(x) \cap Q$, where Q is the polyhedron (invariant under the translations by $\mathbf{Z}^s$) $\mathbf{R}^s \times [0,1]^{n-s}$ in $\mathbf{R}^n$ (Figure 4.9.4).

Note that the translates of P by $\mathbf{Z}^s$ form a partition of Q because we have defined $P = [0,1[^n$ and not $[0,1]^n$. This choice implies that the above result, namely $\sum_\alpha M(x - \alpha|X)$ constant, is true again if $n = s$, which would not be the case if we had set $P = [0,1]^n$.

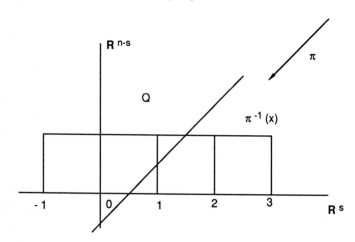

Figure 4.9.4 (s = 1)

This completes the proof of Lemma 4.9.8.

4.9.9 Remarks 1) In general one "normalizes" the box-splines $M(x|X)$ by multiplying them by a constant, so that one obtains a partition of unity: $\sum M(x - \alpha|X) = 1$, as for the one variable uniform B-splines.

2) With the above notation, and denoting by $\mathbf{P}$ the space of all polynomials, one can prove the following equality: $S_X \cap \mathbf{P} = \mathcal{D}(X)$.

3) The fact that S_X contains the polynomials of degree $\leq d$ implies classically that one can approximate any function $f \in L_p(\mathbf{R}^s)$ of class $\mathcal{C}^\infty$, up to the order h^{d+1} by elements of the form $g(x/h)$, $g \in S_X$.

Let us now consider the problem of the linear independence (in the space S_X) of the translate functions $M(x - \alpha|X)$ $(\alpha \in \mathbf{Z}^s)$. We will say that the functions $M(x - \alpha|X)$ $(\alpha \in \mathbf{Z}^s)$ are independent if any relation $\sum_{\alpha \in \mathbf{Z}^s} \lambda_\alpha M(x - \alpha|X) = 0$ implies $\lambda_\alpha = 0$. This notion of independence is stronger that the classical one, since it permits possibly infinite linear combinations, which are meaningful since the $M(x - \alpha|X)$ form a locally finite family.

4.9.10 Proposition *With the preceding notation and hypotheses, the functions $M(x - \alpha|X)$ $(\alpha \in \mathbf{Z}^s)$ are linearly independent in the above sense if and only if any subset of X with s elements forming a basis of $\mathbf{R}^s$ as a vector space is in fact a basis of $\mathbf{Z}^s$ as an abelian group. This condition is equivalent to the fact that for every free family $(X_{i_1}, \ldots, X_{i_s}) \subset X$, $|det(X_{i_j})| = 1$.*

Proof Recall that the elements X_i of X have integer coordinates, and that we assume that X generates $\mathbf{R}^s$.

a) *Necessity.* Let us consider a free family $W = (X_{i_1}, \ldots, X_{i_s}) \subset X$ such that $|det(X_{i_j})| > 1$. The elements of W generate then (over $\mathbf{Z}$) a sublattice Λ of $\mathbf{Z}^s$, and we have

$$\sum_{\alpha \in \Lambda} M(x - \alpha|W) = \frac{1}{|detW|}$$

by 4.8.4, c) since then the supports of the functions $M(x - \alpha|W)$ are disjoint and form a partition of $\mathbf{R}^s$. But if k is an element of $\mathbf{Z}^s \setminus \Lambda$, we have in the same manner

$$\sum_{\alpha \in \Lambda} M(x - \alpha - k)|W) = \frac{1}{|detW|} \, ,$$

which implies the relation

$$\sum_{\alpha \in \Lambda} M(x - \alpha|W) - \sum_{\alpha \in \Lambda} M(x - \alpha - k)|W) = 0.$$

b) *Sufficiency.* We will argue by induction on $n = \text{card } X$, always with the hypothesis that X generates $\mathbf{R}^s$.

If $n = s$, the assertion is clear, because the supports of the functions $M(x - \alpha|X)$, $\alpha \in \mathbf{Z}^s$, are disjoints (it is here where we use the hypothesis).

If $n > s$, let us consider the relation:

$$\sum_{\alpha \in \mathbf{Z}^s} \lambda_\alpha M(x - \alpha|X) = 0.$$

If X_i is an element of X, we deduce from 4.9.4:

$$\sum_{\alpha \in \mathbf{Z}^s} (\Delta_{X_i} \lambda_\alpha) M(x - \alpha|X \setminus X_i) = 0$$

with $\Delta_{X_i} \lambda_\alpha = \lambda_{\alpha + X_i} - \lambda_\alpha$.

i) If $X \setminus X_i$ generates $\mathbf{R}^s$, the induction hypothesis implies that $\Delta_{X_i} \lambda_\alpha = 0$ for all $\alpha \in \mathbf{Z}^s$.

ii) If $X \setminus X_i$ does not generate $\mathbf{R}^s$, $X \setminus X_i$ generates a hyperplane H which we may identify with $\mathbf{R}^{s-1}$, and $M(x|X \setminus X_i)$ is then a box-spline on $\mathbf{R}^{s-1}$, to which we can apply the induction hypothesis (see Remark 4.9.3,b).

In all the cases, we then have $\Delta_{X_i} \lambda_\alpha = 0$ for $\alpha \in \mathbf{Z}^s$ since this is true for any X_i in X, and as X generates $\mathbf{R}^s$, we deduce $\lambda_\alpha = 0 \quad \forall \alpha \in \mathbf{Z}^s$.

∎

5

Triangulations

INTRODUCTION

In this chapter, we will study the triangulations of subsets of the plane, and in particular, with n points of $\mathbf{R}^2$ known, give methods for finding a triangulation of their convex hull with these points as vertices, which is "as best as possible". This is useful in the finite elements method, and also in CAD. For instance, consider a surface $S \subset \mathbf{R}^3$ defined by a family of points Q_i which project on points P_i of $\mathbf{R}^2$: one must then triangulate the convex hull of those points to obtain a piecewise linear approximation of S, or to construct triangular Bézier patches with control points Q_i.

We will consider triangulations of "Delaunay" type, and their definition will be extendable to $\mathbf{R}^3$ (or even $\mathbf{R}^k$ with $k \geq 3$).

Thus we are going to be introduced to computational geometry, which, although an area of recent study, is expanding rapidly.

We will generally assume that we are in $\mathbf{R}^k$ (for any integer k), assuming $k = 2$ only when indispensable, or to simplify the presentation.

5.1 VORONOI DIAGRAMS

Let, $(P_i)_{0 \le i \le n}$ be a set of points in $\mathbf{R}^k$ $(k \ge 2)$ in *"general position"* (this means here that there are no $k + 2$ points on the same hyperplane).

Let $H(P_i, P_j)$ be the closed half-space containing P_i, with boundary the bissector hyperplane of the segment P_iP_j: see Figure 5.1.1.

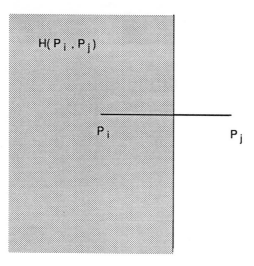

Figure 5.1.1

Set:

$$V(P_i) = V_i = \bigcap_{j \ne i} H(P_i, P_j).$$

5.1.1 Definition   *The set formed by the $V_i's$ $(0 \le i \le n)$ and their mutual intersections is called the Voronoï diagram associated to the points P_i; V_i is the set of points $x \in \mathbf{R}^k$ such that $d(x, P_i) \le d(x, P_j)$ $\forall j \ne i$.*

We will denote by $[P_0 \ldots P_n]$ the convex hull of the points $P_0, \ldots, P_n$.

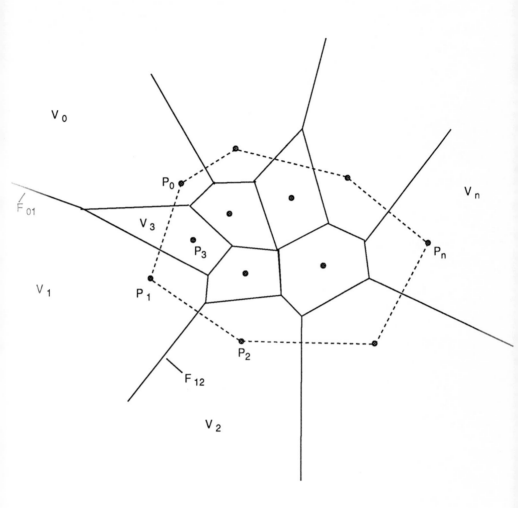

Figure 5.1.2

Figure 5.1.2 shows a Voronoï diagram in the plane. The dotted polygon is the boundary of the convex hull of the P_i's.

5.1.2 Proposition

1) V_i is a convex polytope of dimension k, containing P_i in its interior.

2) Each vertex of V_i is in $k+1$ polytopes V_j, and in $k+1$ edges of dimension 1.

3) The polytope V_i, is unbounded if and only if P_i is in the boundary of the convex hull $[P_0 \dots P_n]$ of the P_j's.

4) If P_s is one of the nearest points of P_i (among the $P_j's$, $j \neq i$), one of the faces of V_i is contained in the mediator hyperplane of P_iP_s.

5)

$$\bigcup_{i=1}^{n} V_i = \mathbf{R}^k, \quad \overset{\circ}{V_i} \cap \overset{\circ}{V_j} = \emptyset \quad \text{for } i \neq j$$

($\overset{\circ}{V_i}$ denotes the interior of the polytope V_i).

Remark   We will see that the hypothesis of general position is used only in condition 2).

Proof of 5.1.2

1) Clear, since V_i is the intersection of $n-1$ closed half-spaces which are convex sets containing P_i in their interior.

2) Set $F_{i,j} = V_i \cap V_j$, and consider a vertex S of the Voronoï diagram; there then exist r faces $F_{i,j}$ of dimension $n-1$ such that $S \in \bigcap F_{i,j}$, and we have $r \geq k$, since the intersection of p of those faces is of codimension $\geq p$, by the dimension formula from linear algebra. One may assume that the points P_i are labeled in such a way that the r faces are $F_{0,1}, F_{1,2}, \ldots, F_{r-1,r}$.

Each $F_{i,j}$ is by definition the set of points $x \in \mathbf{R}^k$ such that

$$d(x, P_i) = d(x, P_j) \leq d(x, P_s) \text{ for all } s \quad (0 \leq s \leq n);$$

we have therefore

$$d(S, P_0) = \cdots = d(S, P_r) < d(S, P_s) \quad \text{for all } s \neq 0, \ldots, r,$$

and S is the center of a sphere passing through $P_0, \ldots, P_r$, and containing no other point P_s. The general position hypothesis then implies that $r \leq k$, and therefore that $r = k$, since we saw above that we also had $r \geq k$. Thus

$$S \in V_0 \cap \ldots \cap V_k, \quad S \notin V_s \quad \text{for} \quad s \neq (0, \ldots, k).$$

Moreover, the edges of V of dimension 1 passing through S are the intersections of k V_i's among $V_0, \ldots, V_k$: there are therefore $k+1$ of them also.

3) We prove that the following conditions are equivalent.
a) V_i is bounded.

b) P_i is not in the boundary of the convex hull C of the P_j's.

Recall that a point P is in the boundary ∂C of C if and only if there exists a supporting hyperplane H passing through P, i.e., such that the P_i's are all on the same side of H (Figure 5.1.3).

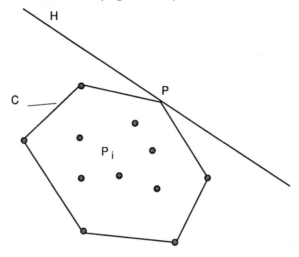

Figure 5.1.3

Proof of a) $\Longrightarrow$ *b)* Let H be a hyperplane passing through P_i and D a half-line passing through P_i and not contained in H (see Figure 5.1.4); V_i being bounded by assumption, D necessarily meets a face F_{ij} of V_i at a point $y \in V_i \cap V_j$. Therefore y is in the mediator hyperplane of P_iP_j, which implies that P_j is on the same side of H as D. This implies that H cannot be a supporting hyperplane.

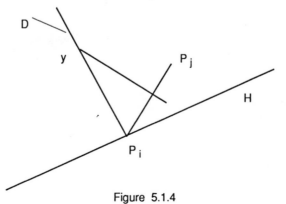

Figure 5.1.4

Proof of b) $\implies$ *a)* Let $P_i \notin \partial C$ and D a half-line passing through P_i; we have to show that one cannot have $D \subset V_i$. Let H be the hyperplane passing through P_i and orthogonal to D (see Figure 5.1.4); H is not a supporting hyperplane by hypothesis, therefore there exists a point P_j on the same side of H than D, whence $y \in D \cap V_j$ (see Figure 5.1.4), and $D \not\subset V_i$.

4) Let y be the middle of $P_i P_s$. We have

$$d(y, P_i) = d(y, P_s) \le d(y, P_r) \quad \forall r,$$

and therefore $y \in V_i \cap V_s$.

5) It is enough to note that we have

$$V_i = \{x \in \mathbf{R}^k \mid d(x, P_i) \le d(x, P_j) \quad \forall j\},$$

and

$$\overset{\circ}{V_i} = \{x \in \mathbf{R}^k \mid d(x, P_i) < d(x, P_j) \quad \forall j \ne i\}.$$

5.2 GENERALITIES ABOUT TRIANGULATIONS

a) Case of the plane $\mathbf{R}^2$

Let (T_i) be a finite family of (closed) triangles, i.e., simplices of dimension 2, $E = \bigcup T_i$.

One says that the T_i's form a *triangulation* of E, if $\overset{\circ}{T_i} \cap \overset{\circ}{T_j} = \emptyset$ for $i \ne j$ (here $\overset{\circ}{T_i}$ denotes the interior of the simplex T_i), and if for any couple (i,j), $i \ne j$, the intersection $T_i \cap T_j$ is either empty, a vertex or a common edge.

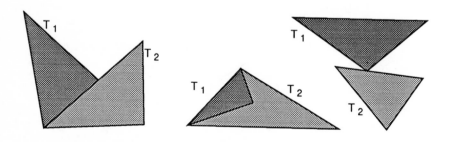

Figure 5.2.1

Three examples of forbidden positions for a triangulation

One says that a polygon is simple if it is made of a closed polygonal line without crossing.

5.2.1 Proposition   *Let E be the interior of a simple polygon (with its boundary), (T) a triangulation of E, v the number of vertices of (T), e the number of edges, f the number of faces (or triangles). We have then*

$$(5.2.1, 1) \qquad\qquad f - e + v = 1.$$

This is known as the "Euler relation".

    Proof   Let us argue by induction on the number f of triangles, the relation being clearly true for $f = 1$.

    Let E be a polygon triangulated with f triangles. Consider the polygon $E' = E \setminus T_1$, where T_1 is a triangle with at least one side in the boundary of E. The two possibilities (for the triangle T_1) are shown in Figure 5.2.2.

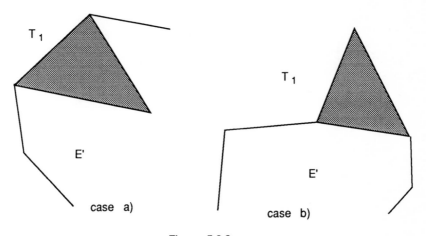

Figure 5.2.2

In both cases, $(5.2.1, 1)$ for E' implies the same one for E; in case a) one deletes a face and an edge, and in case b), one deletes a face, two edges, and one vertex.

∎

5.2.2 Corollary   *Let E be the interior of a simple polygon, triangulated with n vertices, e' edges in the boundary, e'' edges not in the boundary (i.e., common to two triangles), and f faces. We then have*

$$(5.2.2, 1) \qquad\qquad f = 2n - 2 - e' .$$

Proof From (5.2.1, 1) we have $f - (e' + e'') + n = 1$; each face has 3 edges, and the interior edges are common to two faces, whence $3f = e' + 2e''$, and (5.2.2, 1) follows, eliminating e'' between these two relations.

∎

5.2.3 Remarks

1) 5.2.1 is not necessarily true if E is not the interior of a simple polygon. For instance, if E is a "crown" (see Figure 5.2.3), one finds $f - e + v = 0$. More generally, the quantity $\chi(E) = f - e + v$ is called the *Euler-Poincaré characteristic* of E, and does not depend on the chosen triangulation. For further information, the reader should refer to a book on algebraic topology, [M] or [G] for instance.

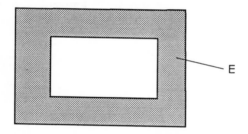

Figure 5.2.3 : $\chi(E) = 0$

2) Corollary 5.2.2 implies that if we fix n points P_i, the number f of triangles we need for the triangulation of the convex hull of the P_i's, the set of vertices of the triangulation being the set of the P_i's, is fixed and given by (5.2.2, 1); we will see that this is not the case in $\mathbf{R}^3$.

b) Case of a surface

Consider a surface $S \subset \mathbf{R}^n$ and $f : U \longrightarrow S$ an injective parametrization of a part, or a chart, of S with for instance $U = [0,1] \times [0,1] \subset \mathbf{R}^2$. Then the image by f of a triangulation of U is called triangulation of $f(U)$. If S is compact, a triangulation of S may be obtained by gluing the triangulations of several charts. Note that the edges and faces are no longer linear.

One may in the same way as above compute $\chi(S)$ and prove that this number is independent of the chosen triangulation (see [M] or [G]).

5.2.4 Example Let S^2 be the two-dimensional sphere, T^2 the torus (homeomorphic to $S^1 \times S^1$). We have then $\chi(S^2) = 2$, $\chi(T^2) = 0$. For the proof, it is enough to construct a triangulation and to count the number of faces, edges and vertices.

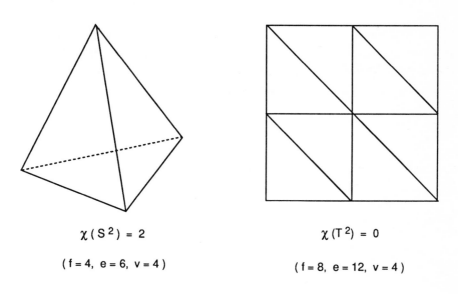

$\chi(S^2) = 2$

$(f = 4,\ e = 6,\ v = 4)$

$\chi(T^2) = 0$

$(f = 8,\ e = 12,\ v = 4)$

Figure 5.2.4

For T^2, the opposite sides of the square are identified, as is suggested by the arrows of Figure 5.2.5. In particular, the four vertices of the square correspond to an unique point M of the torus (see Figure 5.2.5).

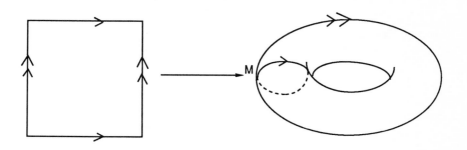

Figure 5.2.5

c) Case of a polytope in $\mathbf{R}^3$

The simplices in dimension three are tetrahedra, and for a triangulation, the intersection of two of them must be either a common face, or a common edge, or a common vertex.

5.2.5 Proposition   Let $E \subset \mathbf{R}^3$ be a polytope homeomorphic to a ball, T a triangulation of the interior of E (and of its boundary ∂E). Assume that T has t tetrahedra, f faces (triangles), e edges and v vertices. We have then the relation (again called Euler's relation)

$$(5.2.5, 1) \qquad\qquad v - e + f - t = 1 .$$

The proof, by induction on t, is similar to that of 5.2.1 and left to the reader.

5.2.6 Remarks

1) From 5.2.4, we see that the triangulation of the boundary ∂E of E satisfies the relation $v' - e' + f' = 2$, if ∂E has v' vertices, e' edges and f' faces.

But we have $3f' = 2e'$, since each edge is the intersection of two faces, and each face is a triangle; we have therefore $f' = 2v' - 4$ and $e' = 3v' - 6$.

2) "On the number of tetrahedra"

Set $v'' = v - v'$, where v'' is the number of vertices in the interior of the polytope, and $f'' = f - f'$. We have $t = v - e + f - 1$ by (5.2.5,1), and $4t = 2f'' + f' = 2f - f'$; each interior face is the intersection of two tetrahedra, and each tetrahedron has 4 faces. We may therefore eliminate f between the two relations

$$\begin{cases} t = v - e + f - 1 \\ 4t = 2f - (2v' - 4), \end{cases}$$

whence

$$t = e - v - v' + 3 = e - 2v' - v'' + 3.$$

But we have clearly $e \leq \binom{v}{2} = \frac{v(v-1)}{2}$, whence $t \leq \frac{v(v-1)}{2} - v - v' + 3 = \binom{v-1}{2} - v' + 2$. Moreover, we have $2e'' \geq 4v''$ since each interior vertex is the intersection of at least four edges, whence

$$t = e - 2v' - v'' + 3 = e'' + v' - v'' - 3 \geq v - 3.$$

Therefore, if we fix the number of vertices v of a triangulation of the interior of a polytope homeomorphic to a ball, the number t of tetrahedra is not fixed, but satisfies

$$(5.2.6, 1) \qquad v - 3 \le t \le \binom{v-1}{2}.$$

Note that the two bounds are equal for $t = 1$. If $v'' = 0$, we have $t = f' = 2v - 4$, but in general, the number of tetrahedra is not a linear function of v.

For more details about these questions, see [E-P-W].

5.3 DELAUNAY TRIANGULATIONS

Let $(P_i)_{0 \le i \le n}$ be a family of points in $\mathbf{R}^k$ (for practical purposes, we will assume $k = 2$ or $k = 3$, but all the results of this section work for any k). We will consider triangulations of the convex hull $[P_i]$ of the points P_i.

5.3.1 Definition   Let P_i $(0 \le i \le n)$ be points in $\mathbf{R}^k$; a triangulation T of the convex hull $[P_i]$ of the P_i's is said to be associated to the points P_i if the following conditions are fulfilled:

a) *The vertices of each simplex of T are taken among the points P_i.*

b) *Every point P_i $(0 \le i \le n)$ is a vertex of at least one simplex of T.*

5.3.2 Proposition   With the above notation, assume $k = 2$, and consider a family T of triangles whose vertices are taken among the points P_i; then this family constitutes a triangulation associated to the points P_i if and only if the following conditions are fulfilled:

a) *Each triangle $P_i P_j P_k$ of T contains no point P_l in its interior.*

b) *If $P_i P_j$ is an edge of the boundary of the convex hull of the P_i's, it is the side of a unique triangle of T.*

c) *If $P_i P_j$ is the side of a triangle of T, and is not an edge of the boundary of the convex hull, then $P_i P_j$ is common to exactly two triangles of T.*

d) *Each P_i is a vertex of at least one triangle of T.*

Proof   The proof is easy, and left to the reader (these properties can be taken as definition of a triangulation associated to the points P_i); the only thing to check is that the union of the triangles of T is equal to $[P_i]$, which follows from d), b) and c).

5.3.3 Definition   The Delaunay triangulation of the convex hull $[P_i]$ of the points P_i is formed from the set $\mathcal{D}$ of the simplices whose vertices are taken among the P_i's, obtained by duality from the Voronoi diagram (see 5.1.1). This means that

P_iP_j is an edge of $\mathcal{D}$ if and only if $V_i \cap V_j \neq \emptyset$;

$P_iP_jP_k$ is a face of $\mathcal{D}$ if and only if $V_i \cap V_j \cap V_k \neq \emptyset$;

etc...

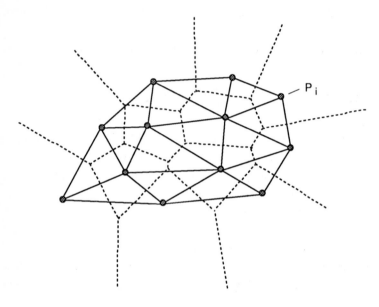

Figure 5.3.1

Delaunay triangulation. Dotted lines represent Voronoi diagram.

5.3.4 Proposition   With the above notation, assume $n \geq k$ and the points $(P_i)_{0 \leq i \leq n}$ in general position in $\mathbf{R}^k$. That is, assume that among the P_i's, there are no $k + 2$ of them which are on a same sphere. Then

1) $\mathcal{D}$ is a triangulation associated to the points P_i (see 4.3.1).

2) It is characterized, among the triangulations associated to the points P_i, by the following property: each sphere, of dimension $k - 1$, circumscribed to a simplex of $\mathcal{D}$ contains no point P_l in its interior.

Proof   For simplicity, we will assume that we are in the plane, i.e., that $k = 2$; the generalization to the general case $k \geq 3$ does not introduce any particular difficulty, and is left to the reader as an exercise.

1) Since each vertex of the Voronoï diagram V is the intersection of three sets V_j (see 5.1.2, 2, this results from the general position hypothesis), one may deduce that each face of $\mathcal{D}$ is a *triangle*. If, without the general position hypothesis, four points, for instance P_1, P_2, P_3, P_4, were cocyclic, they would form a face of $\mathcal{D}$. To obtain a triangulation, one should then subdivide in triangles the non-triangular faces of $\mathcal{D}$.

Let us now prove that the union of these triangles form a triangulation associated to the points P_i. Since it is clear that the vertices of the triangles are taken among the P_i's, it is enough to check the conditions 5.3.2.

a) Let $P_i P_j P_k$ be a triangle of $\mathcal{D}$; it corresponds to a vertex S of the Voronoi diagram V such that $\{S\} = V_i \cap V_j \cap V_k$. S is then equidistant from P_i, P_j and P_k; moreover the circle with center S passing through P_i, P_j and P_k does not contain any other point P_l in its interior, by definition of V, since we have $d(S, P_i) = d(S, P_j) = d(S, P_k) < d(S, P_l)$, for $l \neq i$, j or k (see 5.1.1); in particular, the triangle $P_i P_j P_k$ does not contain any other point P_l.

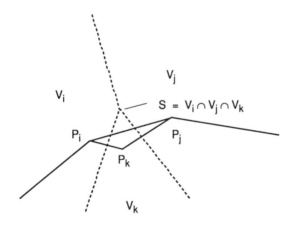

Figure 5.3.2

b) Let $P_i P_j$ be an edge of the boundary of the convex hull of the P_i's; then V_i and V_j are not bounded (see 5.1.2, 3)), and in the case where $n \geq 3$, $V_i \cap V_j$ is a half-line contained in the mediatrix of $P_i P_j$, and containing therefore a unique vertex S of V. This proves that $P_i P_j$ is the side of one and only one triangle of $\mathcal{D}$, since by duality, the triangles of $\mathcal{D}$ with one side equal to $P_i P_j$ correspond to the vertices of V which are on $V_i \cap V_j$ (see Figure 5.3.2).

c) Let P_iP_j be an edge of $\mathcal{D}$ not in the boundary of the convex hull $C(P_i)$; a segment of the mediatrix is then in $V_i \cap V_j$, and this segment is bounded by two vertices of the Voronoi diagram (see 5.1.2, 3)), which correspond to two triangles having a common edge.

d) Consider a point P_i. Since V_i is non empty, there exists j such that $V_i \cap V_j \neq \emptyset$ because $n \geq k$ and $\mathbf{R}^2 = \bigcup_i V_i$; P_iP_j is therefore by definition an edge of a triangle of $\mathcal{D}$ with P_i and P_j as two vertices.

2) We have shown in a) above that the triangulation $\mathcal{D}$ satisfies the given property 5.3.4, 2. It is then clear that this property characterizes $\mathcal{D}$, because if P_i, P_j, P_k are three points such that the circumscribed circle to $P_iP_jP_k$ does not contain any other point P_l, $S = V_i \cap V_j \cap V_k$ is then by definition a vertex of the Voronoi diagram, and therefore $P_iP_jP_k$ is a triangle of $\mathcal{D}$.

∎

5.3.5 Remark   The property 5.1.2, 4) of the Voronoï diagram implies that if P_s is one of the closest points of P_i (among the points $P_0, \ldots, P_n$), the segment P_iP_s is then an edge of $\mathcal{D}$.

We will now prove that in the plane, the Delaunay triangulation has the following remarkable property: it maximises the lower bound of the angles of the triangles (for all the triangulations associated to the P_i's).

5.3.6 Lemma   *Let A,B,C,D be four points non cocyclic in the plane, forming a convex quadrangle. Then:*

a) There are two triangulations of the convex hull $[A, B, C, D]$, the Delaunay triangulation $\mathcal{D}$, and a triangulation $\mathcal{D}'$. One passes from $\mathcal{D}$ to $\mathcal{D}'$ by "exchanging diagonals".

b) The triangulation $\mathcal{D}$ is the one which maximises the lower bound of the angles of the triangles.

Proof   It is clear that there are only two triangulations of the quadrangle $ABCD$; if the circumscribed circle to ABD for instance does not contain C, the triangulation $\mathcal{D}$ (which exists by 5.3.4) is the one containing the edge BD, and the other diagonal AC is an edge of the triangulation $\mathcal{D}'$ (since then the circumscribed circle to ABC contains D).

We see moreover, by considerations of elementary geometry that every angle of a triangle of $\mathcal{D}$ is greater or equal to an angle of a triangle of $\mathcal{D}'$.

For instance, in the case of Figure 5.3.3, we have $\widehat{BA, BD} > \widehat{CA, CD}$, since the circumscribed circle to the triangle ABD does not contain C.

∎

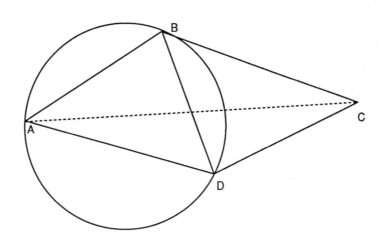

Figure 5.3.3

In the Figure 5.3.3, the edges of the triangulation $\mathcal{D}$ are in solid lines, and one edge of $\mathcal{D}'$ is a dotted line.

5.3.7 Definition   *Assume the points $(P_i)_{0 \leq i \leq n}$ in general position in the plane. A triangulation associated to the P_i's is called locally Delaunay if its restriction to any pair of triangles having a common edge is a Delaunay triangulation (see 5.3.6).*

5.3.8 Proposition   *Let $\mathcal{T}$ be a triangulation associated to the points $P_i \in \mathbf{R}^2$ ($0 \leq i \leq n$) which we assume in general position. The following conditions are equivalent:*

    a) $\mathcal{T}$ is the Delaunay triangulation

    b) $\mathcal{T}$ is a triangulation locally Delaunay (see 5.3.7).

Proof

    $a) \Longrightarrow b)$ Clear by 5.3.4, 2).

    $b) \Longrightarrow a)$ Let $\mathcal{D}$ be the Delaunay triangulation, $\mathcal{T}$ a triangulation verifying b).

Let us first show that to prove that $\mathcal{D} = \mathcal{T}$, it is enough to verify that any triangle of $\mathcal{T}$ having at least one common edge with a triangle of $\mathcal{D}$ is in fact a triangle of $\mathcal{D}$.

Let E be the set of triangles of T which are not triangles of $\mathcal{D}$; since this set is finite, if it is not empty, it contains at least one triangle having an edge belonging to only one triangle of E. This edge is then necessarily common to a triangle of E and to a triangle of $\mathcal{D}$ (this results immediately from the fact that $\mathcal{D}$ and T are triangulations associated to the points P_i; in particular, the edges of the convex hull $[P_0 \ldots P_n]$ are common edges to a triangle of $\mathcal{D}$ and to a triangle of T).

We may therefore only consider a triangle of T having a common edge with a triangle of $\mathcal{D}$.

Let then $P_1 P_2$ be a common edge to T and $\mathcal{D}$, and consider a triangle $P_1 P_2 P_3$ of T. We will assume that this triangle does not belong to $\mathcal{D}$, and prove this is absurd.

There exists then a point R among the P_i's such that the triangle $P_1 P_2 R$ is a triangle of $\mathcal{D}$, and that R is on the same side as P_3 with respect to the line $P_1 P_2$. This point R is then in the circumscribed circle C to $P_1 P_2 P_3$, since, as $\mathcal{D}$ has property 5.3.4, 2), the circumscribed circle to $P_1 P_2 R$ cannot contain the point P_3 (see Figure 5.3.4).

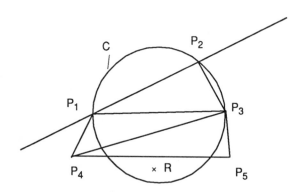

Figure 5.3.4

On the contrary, R is not in the triangle $P_1 P_2 P_3$, by definition of a triangulation associated to the P_i's, use 5.3.2, a), applied to T.

Assume for instance that R and P_2 are on each side of the line $P_1 P_3$ (otherwise we would echange the roles of P_1 and P_2), and let P_4 be the point $P_i \neq P_2$ such that the triangle $P_1 P_3 P_4$ be a triangle of T (P_4 exists, because $P_1 P_3$ is not an edge of the boundary of the convex hull of the P_i's, and then $P_1 P_3$ is common to two triangles of T; see 5.3.2, c)).

We deduce that P_4 is not in the interior of the circle C (because T is locally Delaunay), and then $P_4 \neq R$, so R is not in the triangle $P_1 P_3 P_4$. For otherwise, the circumscribed circle to $P_1 P_3 P_4$ contains R, since it contains the circle arc of C bounded by P_1 and P_3, and not containing P_2.

The situation of R with respect to the triangle $P_1 P_3 P_4$ is similar to its situation with respect to $P_1 P_2 P_3$: we may then repeat the same operation with a point $P_5 \neq R$ such that $P_3 P_4 P_5$ is a triangle of T (if P_1 and R are on each side of $P_3 P_4$; otherwise, we would consider the triangle $P_1 P_4 P_5$), and the situation of R with respect to $P_3 P_4 P_5$ would be again the same (i.e., R not in the triangle $P_3 P_4 P_5$, and R in the circumscribed circle to $P_3 P_4 P_5$). One may then repeat indefinitely the same operation without ever reaching the point R, which is in contradiction with the fact that the P_i's are finite in number.

∎

5.3.9 Corollary   *Among all the triangulations associated to the P_i's, the Delaunay triangulation $\mathcal{D}$ realizes the maximum of the lower bound of the angles of the triangles.*

Proof   Let T be a triangulation associated to the P_i's. For each triangle t_i of T, let α_i be the measure of the smallest angle of t_i.

If T is not the Delaunay triangulation $\mathcal{D}$, it is not locally Delaunay by 5.3.8, and there exist two triangles of T having a common side which do not make a Delaunay triangulaion of their union. We obtain a Delaunauy triangulation of their union by "exchanging diagonals" (see 5.3.6); doing this, we obtain a new triangulation T', again associated to the P_i's, such that $\sum \alpha_i' > \sum \alpha_i$, the sum being taken for all the triangles of T' (resp. for all the triangles of T); see 5.3.6,b). But since there are only a finite number of possible triangulations associated to the P_i's, there are only a finite number of possible values for the sum $\sum \alpha_i$, therefore the previous process stops after a finite number of steps, giving a triangulation $\mathcal{D}$ locally Delaunay, therefore Delaunay by 5.3.8, and such that every angle of a triangle of $\mathcal{D}$ is greater or equal to an angle of a triangle of T.

∎

5.3.10 Remarks

1) The above process shows that starting with a triangulation associated to the P_i's, one may find $\mathcal{D}$ in a finite number of steps, exchanging successively the diagonals of the convex quadrangle not locally Delaunay.

This may easily be put in the form of an algorithm. In the next section, we will describe another algorithm to construct $\mathcal{D}$, which will have two advantages over this one:

a) It will not assume the knowledge of a triangulation associated to the P_i's.

b) It will immediately generalize to any dimension, which is not the case for above process.

2) The property of reaching the maximum of the lower bound of the angles of the triangles (among the triangulations associated to the P_i's) does not characterize the triangulation $\mathcal{D}$ (i.e., other triangulations associated to the P_i's may have the same property).

3) The triangulation $\mathcal{D}$ does not in general reach the minimum of the upper bounds of the angles of the triangles, among the triangulations associated to the P_i's. Consider for instance Figure 5.3.5 below.

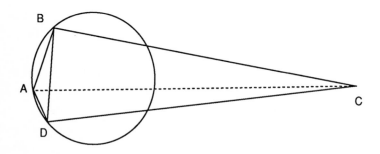

Figure 5.3.5

The triangulation of the quadrangle $ABCD$ having edge BD is the Delaunay triangulation, but the angle $\widehat{AD,AB}$ is greater than the angle $\widehat{DC,DA}$ in particular, and therefore greater than all the angles of the triangles of the other triangulation of $ABCD$.

5.4 CONSTRUCTION ALGORITHM

The following algorithm comes from [H].

Let us first describe the general frame of the algorithm.

a) We start with three points P_0, P_1, P_2 among the P_i's, and we "create" the triangle $P_0 P_1 P_2$. It is the only triangulation $\mathcal{D}_2$ associated to $P_0 P_1 P_2$, and it is clearly Delaunay.

b) Assume we found the Delaunay triangulation $\mathcal{D}_{n-1}$ associated to $P_0, \ldots, P_{n-1}$. We built the Delaunay triangulation $\mathcal{D}_n$ associated to $P_0, \ldots, P_n$. We will describe this construction in the case of the plane for the sake of simplicity. It consists of "destroying" some triangles of $\mathcal{D}_{n-1}$, and "creating" new triangles, in particular triangles with a vertex at P_n.

There are three cases to consider:

1) P_n *is in the convex hull* $C(P_0, \ldots, P_{n-1})$

This is the most important case, because one often assumes be the relevant one, adding to the P_i's three more points P, Q, R, such that the triangle PQR contains all the points P_i (or may be a rectangular box decomposed into two triangles), and starting the algorithm with the triangle PQR. (The obtained result, which is not strictly speaking a triangulation associated to the P_i's, is in general sufficient).

Let Σ be the set of triangles of $\mathcal{D}_{n-1}$ whose circumscribed circle contains P_n (this set is not empty, since P_n is by assumption in the interior of a triangle of $\mathcal{D}_{n-1}$).

Let moreover $(e_j)_{1 \leq j \leq k}$ be the set of edges of the triangles of Σ, not common to two triangles of Σ.

The construction of $\mathcal{D}_n$ consists then of:

    a) Destroying the triangles of Σ.

    b) Creating the triangles $P_n e_j$ $(1 \leq j \leq k)$.

In other words, one sets:

$$\mathcal{D}_n = (\mathcal{D}_{n-1} \setminus \Sigma) \bigcup \left\{ P_n e_j \right\}.$$

α) *Proof that* $\mathcal{D}_n$ *is a triangulation associated to the* P_i's

Let V denote the union of the triangles of Σ. We have $P_n \in V$ since, by hypothesis P_n is inside a triangle of $\mathcal{D}_{n-1}$.

5.4.1 Lemma

a) V is a starred domain with respect to the point P_n, with boundary $\bigcup_j a_j$.

b) Every vertex of a triangle of Σ belongs to an edge e_j.

Proof a) Let $e_j = AB$ be an edge of the boundary of V. If γ is a point of e_j, we have to prove that the segment $P_n\gamma$ is entirely in V, i.e., cannot cut another edge $e_k = CD$ of the boundary of V (see Figure 5.4.1).

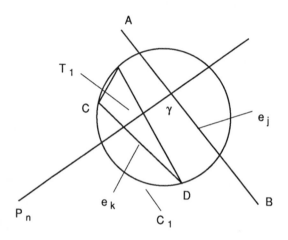

Figure 5.4.1

Assume that the segment $P_n\gamma$ cuts the edge CD of the boundary of V. Then CD is not an edge of the boundary of the convex hull of the P_i's; recall that A, B, C, D are taken among the P_i's. There exists therefore a triangle T_1 of $\mathcal{D}_{n-1} \setminus \Sigma$ having the edge $e_k = CD$, whose circumscribed circle C_1 does not contain P_n. But, by hypothesis, $e_j = AB$ is in the boundary of a triangle T_2 of Σ, whose circumscribed circle C_2 contains P_n; then C_2 contains the segment $P_n\gamma$, and therefore one of the two arcs of the circle C_1 delimited by the line $P_n\gamma$. This implies that C_2 contains either the point C, or the point D, which is incompatible with the hypothesis that $\mathcal{D}_{n-1}$ is a Delaunay triangulation.

Proof of b) Let C be a vertex of a triangle of Σ. If C were not on the boundary of V, it would be in the interior of V; the line P_nC would then cut an edge e_j of the boundary of V. Since e_j is the edge of a triangle of Σ, the circumscribed circle to this triangle would contain P_n, and therefore C, which is absurd.

∎

It is now easy to prove that $\mathcal{D}_n$ is a triangulation associated to the points P_i: the verification of the conditions a),b),c),d) of 5.3.2 is straightforward and left to the reader.

β) Proof that $\mathcal{D}_n$ is the Delaunay triangulation associated to the points $P_0, \ldots, P_n$

It is enough to see that a circle C circumscribed to a triangle of $\mathcal{D}_n$, with vertex P_n does not contain any other point P_k (see 5.3.4; for the circumscribed circles to the triangles of $\mathcal{D}_n$ having no vertex equal to P_n, this is true by construction).

Assume that the circumscribed circle to $P_n AB$ contains a point P_k (see Figure 5.4.2).

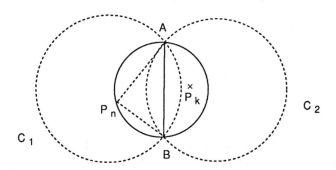

Figure 5.4.2

By hypothesis, AB is an edge of a triangle of Σ, therefore of a triangle of $\mathcal{D}_{n-1}$ whose circumscribed circle C_1 contains P_n (and not P_k, since D_{n-1} is the Delaunay triangulation of $P_0, \ldots, P_{n-1}$). P_k is then placed as on Figure 5.4.2: AB is therefore not an edge of the boundary of the convex hull $\partial C(P_0, \ldots, P_n)$, but is an edge of a second triangle of $\mathcal{D}_{n-1}$ not in Σ, whose circumscribed circle C_2 does not contain P_n. But then C_2 contains P_k, which is incompatible with the hypothesis on $\mathcal{D}_{n-1}$.

∎

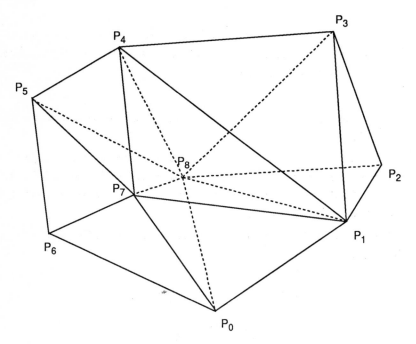

Figure 5.4.3 : case 1) (n = 8)

Figure 5.4.3 illustrates the algorithm in the first case, for $n = 8$; Σ is formed with the five triangles $P_0P_1P_7$, $P_1P_2P_3$, $P_1P_3P_4$, $P_1P_4P_7$, $P_4P_5P_7$. The created triangles are in dotted lines.

For the sake of completeness, let us now consider the two other cases which may occur (but which may be avoided by the process described in 1) above).

2) P_n is not in any circumscribed circle to a triangle of D_{n-1}

We then destroy no triangle of $\mathcal{D}_{n-1}$.

One joins P_n to the vertices of the edges of the boundary of the convex hull of $P_0, \ldots, P_{n-1}$, which support lines separating P_n from the set of the other points, and one creates the triangles containing P_n (as vertex) and these edges.

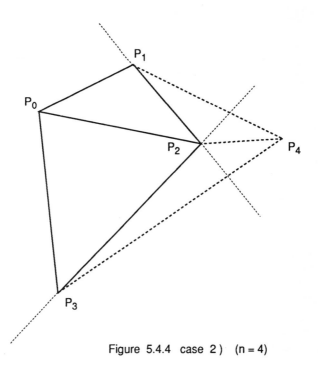

Figure 5.4.4 case 2) (n = 4)

Figure 5.4.4 illustrates the algorithm in the second case ($n = 4$): we keep the triangles of $\mathcal{D}_3$, and we "create" the triangles $P_1 P_2 P_4$ and $P_2 P_3 P_4$.

We have to show the two following assertions:

α) $\mathcal{D}_n$ is a triangulation associated to the points $P_0, \ldots, P_n$.

β) $\mathcal{D}_n$ is the Delaunay triangulation.

Proof of α). It is easy to verify the conditions of Proposition 5.3.2, using the following (immediate) lemma

5.4.2 Lemma

i) An edge $P_i P_j$ is an edge of the boundary of the convex hull of $P_0, \ldots, P_n$ if and only if the line $P_i P_j$ is a supporting line, i.e., if all the points P_i $(0 \leq i \leq n)$ are on the same side of this line.

ii) A point P_i is a point of the boundary of the convex hull $[P_0 \ldots P_n]$ if and only if there exists a supporting line passing through P_i.

Proof of β)

It is enough (after 5.3.8) to verify that $\mathcal{D}_n$ is a triangulation locally Delaunay, which is true by hypothesis for those pairs of triangles one of which is a triangle of $\mathcal{D}_{n-1}$, and immediate for the pairs of triangles created having a common edge P_iP_j (see Figure 5.4.5).

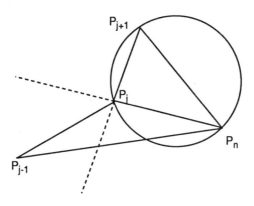

Figure 5.4.5

In Figure 5.4.5, the point P_{j-1} cannot be in the circumscribed circle to $P_jP_{j+1}P_n$, since it necessarily is in the sector delimited by the half-lines in dotted lines containing P_nP_j and $P_{j+1}P_j$, because by construction, the lines containing P_jP_{j+1} and $P_{j-1}P_j$ must separate P_n from the other points P_i.

3) P_n is not in the convex hull $C(P_0, \ldots, P_n)$, but in at least one circumscribed circle to a triangle of D_{n-1}

This situation mixes the cases 1) and 2). Let Σ be the set of triangles whose circumscribed circle contains P_n, $(e_i)_{1 \leq i \leq p}$ the edges of the boundary of the convex hull which are not faces of a triangle of Σ, and separating P_n from the other P_i's. Let also $(b_j)_{1 \leq j \leq q}$ be the edges of the triangles of Σ, not common to two triangles of Σ, and not separating P_n from the others vertices of Σ.

Then the algorithm destroys the triangles of Σ and creates the triangles $\{P_n e_i\}_{0 \leq i \leq p}$ and $\{P_n b_j\}_{1 \leq j \leq q}$. In other words, we have:

$$D_n = (D_{n-1} \setminus \Sigma) \bigcup \{P_n e_i\} \bigcup \{P_n b_j\}$$

5.4.3 Remarks

a) The definition of Σ is the same as for case 1); for case 2), Σ is empty.

b) The definition of the edges e_i is the same as for case 2), except for the edges of the triangles of Σ.

c) The definition of the edges b_j is the same as for case 1), because in case 1), no edge of the boundary of Σ separates P_n from the other points P_i of Σ (see Lemma 5.4.2).

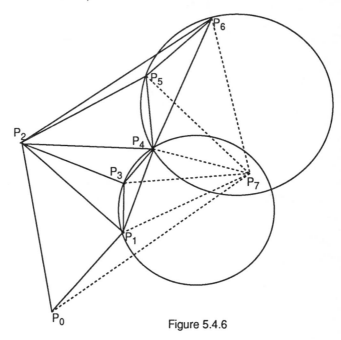

Figure 5.4.6

In Figure 5.4.6 $(n = 7)$, Σ is formed with triangles $P_1P_3P_4$ and $P_4P_5P_6$; there is one edge of type e_i, P_0P_1, and four of type b_j, the edges P_1P_3, P_3P_4, P_4P_5 and P_5P_6. The created edges are in dotted lines.

The proof of the fact that one obtains by this process the Delaunay triangulation associated to $P_0, \ldots P_n$ does not present any difficulty, and is left to the reader.

5.5 REMARKS ABOUT COMPLEXITY

We want to estimate, with respect to n, the maximal number $C(n)$ of operations fulfilled in the construction algorithm of $\mathcal{D}_n$ (this for any possible data of n points $P_0, \ldots, P_n$).

In the frame that we will study, each addition, substraction, multiplication, division or comparison of two real numbers is counted as one operation. We assume therefore that we work with an ideal computer, which is able to compute in an exact way over the real numbers. Even in the case where all the data are rational numbers (or integers), we do not care about the size of the numbers that we use in the computations, and consequently we do not care about the real time we spend on each elementary operation.

We will assume in this section that the situation is planar.

5.5.1 Lemma   *We have*

$$C(n) \geq \Omega(n \log n).$$

This means that there exists a constant $C_1 > 0$ such that $C(n) \geq C_1 n \log n$ for any n.

Proof

a) Note first that, knowing D_n, we deduce (in time linear in n), the Voronoï diagram V_n associated to $P_0, \ldots, P_n$. The vertices of the polygon V_i (associated to P_i: see 5.1.1) are the centers of the circumscribed circles to the triangles of D_n having a vertex at P_i; it is then enough to prove the the complexity $C'(n)$ of the construction of the Voronoï diagram satisfies $C'(n) \geq C_1' n \log n$.

b) Any algorithm which, given n points in general position in the plane, constructs the Voronoï diagram associated to these n points, gives an algorithm letting us sort (in linear time) n real numbers $x_0, \ldots, x_{n-1}$ in increasing order. In fact, given n real numbers $x_0, \ldots, x_{n-1}$, we consider n points of the plane $P_0, \ldots, P_{n-1}$, on the x-axis for instance, with abscissa $x_0, \ldots, x_{n-1}$; (if we want to be in the frame of Section 1, we have to set $P_i = (x_i, y_i)$, the y_i's being very small, and chosen in such a way that the points $P_0, \ldots P_{n-1}$ are in general position).

Knowing the Voronoï diagram, we may then sort the numbers x_i on the following way: given x_{i_0}, we consider among the sides of V_{i_0} that which cuts the x-axis at the closest point of P_{i_0}. If this side is common to V_{i_0} and V_{i_1}, x_{i_1} is the successor of x_{i_0} in the sequence (x_i).

c) Any algorithm which sorts n real numbers has a complexity greater than or equal to $\Omega(n \log n)$ (see [P-S]).

■

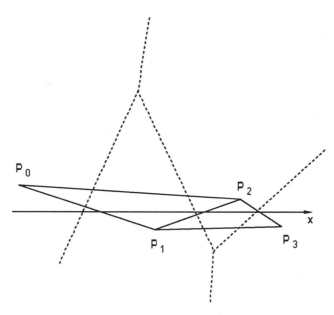

Figure 5.5.1 : x_1 is the successor of x_0

5.5.2 Proposition   *In the case of the plane, the above algorithm has a complexity $O(n^2)$.*

This means that there exists a constant C_2 such that the number of elementary operations fulfilled in the algorithm is less than or equal to $C_2 n^2$. This proposition then has no practical signification, since the constant C_2 is not specified.

Proof   Assume that $\mathcal{D}_{n-1}$ is constructed. Then the search for Σ can be made in time linear in n because the number of triangles of $\mathcal{D}_{n-1}$ is less than or equal to $2(n-1)-5$ by 5.2.2, and then the construction of $\mathcal{D}_n$ from $\mathcal{D}_{n-1}$ can also be made in time linear in n.

The construction of $\mathcal{D}_n$ can therefore be made in time $O(1+\cdots+n) = O(n^2)$.

■

5.5.3 Remarks

1) Staying in the plane, there exists a construction of the Voronoï diagram of n points $(P_i)_{0 \leq i \leq n-1}$ which can be made in time $O(n \log n)$ (see

[P-S]). If we look to Lemma 5.5.1, we see that this process gives in theory an algorithm with an "optimal" complexity for constructing the Delaunay triangulation associated to n points of the plane.

2) One may decompose the $(n-2)^{th}$ step of the above algorithm into two parts; we assume that we are in case number 1:

a) Search for a point of Σ.

P_n being given, let P_k be one of those points nearest to P_n among the points $P_0, \ldots, P_{n-1}$; it follows from Remark 5.3.5 that the segment $P_n P_k$ is an edge of $\mathcal{D}_n$. We deduce that P_k is necessarily a point of Σ (since the segment $P_n P_k$ must pass through a triangle of $\mathcal{D}_{n-1}$ with vertex P_k, and then this triangle must be destroyed).

This part is then reduced to the search for a point nearest to P_n among the points $P_0, \ldots, P_{n-1}$; this search can be made in time $O(\log n)$, if we build, at each step of the algorithm, a "quadtree" (or "octree" if we are in the space) on the P_i's: see [B].

b) When a point P_k of Σ is known, we find a triangle t_1 of Σ (by testing all the triangles with vertex P_k); we find then another triangle t_2 of Σ (if it exists) by testing the three triangles having a side common with t_1. Then we continue by the same process until the test stops.

The fact that Σ is a starred domain (see 5.4.1) shows that we have find all the triangles of Σ.

If, for a configuration of points, the number of triangles of Σ is bounded uniformly in n, part b) above has a complexity bounded by a constant, which implies that the above triangulation algorithm has a complexity of the form $O(n \log n)$. This is for instance the case for configurations of points sufficiently regular (which is in general the case), but not always: one may construct examples where, at each step of the algorithm, all the triangles of the preceding step are in Σ.

6

NOTIONS OF REAL ALGEBRAIC GEOMETRY

This chapter introduces some useful notions for the effective manipulation of polynomials with integer or rational coefficients, and the geometric sets they define. We want for instance answers (in principle exact, without approximation) to questions of the following kind: how many intersection points of two curves are there; what is the equation for the intersection of two surfaces, etc ...

The questions are treated in the polynomial frame, but all the results may apply to spline functions, which are piecewise polynomials, with some care (and supplementary difficulties) due to the fact that in this case we consider only portions of algebraic curves (or surfaces).

This chapter is only an introduction to the subject, so the reader who wants to go further in these matters may look at [D-S-T] for formal computation, and to [B-R] for the geometric notions.

6.1 ROOTS OF ONE-VARIABLE POLYNOMIALS

Let $P \in \mathbf{R}[X]$. We may pose three problems about the roots of P:

a) Determine the exact number of real roots of P.

b) Isolate the roots of P, i.e., find a family of disjoint intervals $[a_i, b_i]$ of $\mathbf{R}$, such that each interval contains one unique root of P.

c) Find an approximate value of each root. This problem needs first the resolution of problem b) above.

In this section, we will study the problems a) and b) without going into deep details; problem c) requires approximation theory beyond the scope of this book.

For more details, the reader may consult [B-R].

a) Number of real roots of a polynomial

6.1.1 Definition   (see 1.5.5). Let $a = (a_0, \dots, a_n)$ be a sequence of real numbers. Then the variation (or number of change of signs) of a, denoted $V(a)$, is the number of pairs $(i, i + k)$, $k \geq 1$ such that

    a) $a_i a_{i+k} < 0$,

    b) $a_{i+r} = 0$ for $0 < r < k$.

6.1.2 Definition   Let $K \subset \mathbf{R}$ be a field, and P and Q be two elements of $K[X]$. Let R be the remainder of the Euclidian division of $P'Q$ by P, P' being the derivative of P. The Sturm sequence of (P, Q), denoted $S(P, Q)$, is the sequence of polynomials S_i $(0 \leq i \leq t)$ such that

$$S_0 = P$$
$$S_1 = R$$
$$S_{i-1} = S_i A_i - S_{i+1}, \quad deg(S_{i+1}) < deg(S_i) \quad (1 \leq i \leq t - 1)$$
$$S_t = (P, R) \quad \text{(the GCD of } P \text{ and } R).$$

In other words, the Sturm sequence is, up to sign, the sequence of remainders of the Euclidian algorithm. Note that if $Q = 1$, we have $R = P'$, and the above notion of Sturm sequence is the classical one.

6.1.3 Proposition   Let $[a, b] \subset \mathbf{R}$ be an interval such that $P(a)P(b) \neq 0$. If $S = S(P, Q)$ denotes the Sturm sequence of P and Q, we have

$$V\big(S(a)\big) - V\big(S(b)\big) = Z_{Q>0}^{[a,b]}(P) - Z_{Q<0}^{[a,b]}(P),$$

$Z_{Q>0}^{[a,b]}(P)$ denoting the number of roots α of P in the interval $[a, b]$ such that $Q(\alpha) > 0$, and $Z_{Q<0}^{[a,b]}(P)$ the number of roots β in $[a, b]$ such that $Q(\beta) < 0$.

Before proving this proposition, let us indicate two useful corollaries. The notation is the same as in Proposition 6.1.3.

6.1.4 Corollary   Let $S_1 = S(P, 1)$ be the Sturm sequence of the polynomials P and 1. Then $V\big(S_1(a)\big) - V\big(S_1(b)\big)$ is equal to the number of roots of P in the interval $[a, b]$.

The proof is immediate, from the proposition.

∎

6.1.5 Corollary   Let $S(+\infty)$ be the sequence of dominating coefficients of the sequence of polynomials $S = (S_i)$ $(0 \leq i \leq t)$, and $S(-\infty)$ be the sequence of dominating coefficients of the sequence $(-1)^{d_i} S_i$, if S_i is of degree d_i. We have then

$$V\big(S(-\infty)\big) - V\big(S(+\infty)\big) = Z_{Q>0}(P) - Z_{Q<0}(P),$$

$Z_{Q>0}(P)$, resp. $Z_{Q<0}(P)$, denoting the number of roots α of P such that $Q(\alpha) > 0$, resp. $Q(\alpha) < 0$.

Proof   This corollary follows from the (immediate) following fact: there exists a number $A > 0$ such that for any $M > A$, we have
a)

$$V\big(S(M)\big) = V\big(S(+\infty)\big)$$
$$V\big(S(-M)\big) = V\big(S(-\infty)\big).$$

b) Any root α of one of the polynomials S_i satisfies $|\alpha| < M$.

∎

Proof of the Proposition
Let $G(X) = S_t(X)$ be the GCD of P and R. G is also the GCD of P and $P'Q$. All the elements of the sequence $S(P,Q)$ are divisible by G, and G is the GCD of two consecutive elements of the sequence $S(P,Q)$. Let S^1 be the sequence of polynomials defined by $P^1 = P/G$, $Q^1 = P'Q/G$, $S_i^1 = S_i/G$ $(2 \leq i \leq t)$. We then have

$$V\big(S^1(a)\big) - V\big(S^1(b)\big) = V\big(S(a)\big) - V\big(S(b)\big).$$

Let us look at the changes in the sequence $S^1(x)$ as x passes from a to b: there is a possible change of value for the sequence $S^1(x)$ only when x passes through a value α, a root of one of the S_i^1's $(0 \leq i \leq t)$.

a) Assume that α is a root of some of the S_i^1's, $0 < i < t$, and that $S_0^1(\alpha) = P^1(\alpha) \neq 0$. We then have, for each index i such that $S_i^1(\alpha) = 0$, $S_{i-1}^1 = S_i^1 A_i - S_{i+1}^1$, which implies $S_{i-1}^1(\alpha)S_{i+1}^1(\alpha) \neq 0$. This is because $S_t^1 = 1$, and S_t^1 is the GCD of S_i^1 and S_{i+1}^1 $(0 \leq i < t)$, so that S_i^1 and S_{i+1}^1 cannot have a common root. This implies that $V\big(S^1(x)\big)$ is constant in a neighborhood of α.

b) Assume $S_0(\alpha) = P(\alpha) = 0$, and set $P(X) = (X - \alpha)^r \tilde{P}(X)$, with $\tilde{P}(\alpha) \neq 0$.

1) If $Q(\alpha) = 0$, $G(X)$, the GCD of P and $P'Q$, is divisible by $(X - \alpha)^r$, which implies $P^1(\alpha) \neq 0$, and we are in case a).

2) If $Q(\alpha) \neq 0$, we have

$$G(X) = (X - \alpha)^{r-1}\tilde{G}(X), \text{ with } \tilde{G}(\alpha) \neq 0,$$
$$P' = r(X - \alpha)^{r-1}\tilde{P}(X) + (X - \alpha)^r \tilde{P}'(X),$$
$$P^1(X) = (X - \alpha)\frac{\tilde{P}(X)}{\tilde{G}(X)},$$
$$Q^1(X) = \frac{P'Q(X)}{G(X)} = \left(r\frac{\tilde{P}(X)}{\tilde{G}(X)} + (X - \alpha)\frac{\tilde{P}'(X)}{\tilde{G}(X)} \right) Q(X).$$

Assume for instance $\dfrac{\tilde{P}(\alpha)}{\tilde{G}(\alpha)} > 0$; the case $\dfrac{\tilde{P}(\alpha)}{\tilde{G}(\alpha)} < 0$ is similar, and treated in the same way.

As by hypothesis $P^1(\alpha) = 0$, Q^1 and R/G have the same (constant) sign on a neighborhood of α; we then have the following sign tables when x is close to α:

		x		α	
$Q(\alpha) > 0:$	P^1	$-$	0	$+$
	$P'Q/G = Q^1$	$+$	$+$	$+$

		x		α	
$Q(\alpha) < 0:$	P^1	$-$	0	$+$
	$P'Q/G = Q^1$	$-$	$-$	$-$

We see that when x "passes the value α", the variation of the sequence $S_i^1(x)$ diminishes by 1 in the case where $Q(\alpha) > 0$, and increases by 1 in the case where $Q(\alpha) < 0$, which achieves the proof of the proposition.

∎

6.1.6 Remark It is possible to multiply (or divide) each term of the Sturm sequence by a positive element of K, without modifying the variation. This is useful when the polynomials P and Q are in $A[X]$, where A is a ring with quotient field K. We may then obtain, using the pseudo-division 6.1.7, a sequence equivalent to the Sturm sequence, but with all the polynomials in $A[X]$ (we generally have $A = \mathbf{Z}$).

6.1.7 Definition   *Let*

$$P = a_p X^p + \cdots + a_1 X + a_0, \quad a_p \neq 0$$
$$Q = b_q X^q + \cdots + b_1 X + b_0; \quad b_q \neq 0$$

be two polynomials in $A[X]$ such that $p \geq q$. The pseudo-division of P by Q is by definition the Euclidian division of $(b_q)^{p-q+1} P$ by Q . The remainder of this division is called pseudo-remainder.

We see immediately that if we make the division:

$$(b_q)^{p-q+1} P = BQ + R$$

with $\deg R < q$, B and R have coefficients in A. Note that it is not necessarily the case for the division of P by Q. The reader may consult [D-S-T], where an algorithm of GCD is described with pseudo-divisions, in such a way that the size of the coefficients of the polynomials increases only linearly with respect to the degrees of P and Q. This algorithm permits us in particular to compute Sturm sequences (for polynomials with coefficients in $\mathbf{Z}$), with a "moderate" increasing of the size of coefficients. It uses the notion of sub-resultant (see Section 6.2).

b) Isolation of roots

We will describe some methods, leading to algorithms, letting us isolate the (real) roots of a polynomial $P \in \mathbf{Z}[X]$.

Isolating roots means finding a sequence $[a_i, b_i]$ of intervals with rational endpoints, such that each interval $[a_i, b_i]$ contains one unique root of P.

We will assume that P is square-free, which implies that all its roots are simple.

We may always assume we are in this case, by dividing P by the GCD of P and of its derivative P'. This GCD may be computed by Euclid's algorithm (see [D-S-T]).

6.1.8 Lemma   *There exists a number m (depending on the degree n of P and of a bound on the size of the coefficients) such that if α_i and α_j are two distinct roots of P, we have $|\alpha_i - \alpha_j| \leq m$.*

It can be proved that $m \geq \exp^{-cn \log(n)}$, where c is a constant > 0: see for instance [B-R].

6.1.9 "Kronecker's method"

The method consists of the following steps

a) Let M be a bound for the $|\alpha_i|$'s, where the α_i's are the roots of P. We then divide the interval $[-M, +M]$ in $2M/m$ intervals (a_i, a_{i+1}) $(1 \leq i \leq 2M/m)$ of length m.

b) We evaluate $P(a_i)$ $(1 \leq i \leq 2M/m)$. If $P(a_i) = 0$, we find a root of P.

c) If $P(a_i) \neq 0$ and $P(a_{i+1}) \neq 0$, we compute $P(a_i)P(a_{i+1})$.

d) If $P(a_i)P(a_{i+1}) < 0$, there is a real root of P in $]a_i, a_{i+1}[$ (and only one, since $|a_i a_{i+1}| \leq m$), and if $P(a_i)P(a_{i+1}) > 0$, there is no root of P in $]a_i, a_{i+1}[$.

This ends the algorithm.

Another method consists of using the properties of Sturm sequences described above.

6.1.10 "Sturm's method"

Let P be a square-free polynomial.

a) We construct a Sturm sequence $S(P, P')$ by Euclid's algorithm (using well choosen pseudo-divisions: see [D-S-T]). If $x \in \mathbf{R}$, let $w(x)$ denote the variation of the sequence $S(P, P')$ at x.

b) Set $x_0 = -M$, $x_1 = M$, $w_1 = w(x_0) - w(x_1)$ and $w := w_1$. If $w = 0$, P has no real root, and if $w = 1$, P has exactly one real root in $[-M, +M]$ (and then also in $\mathbf{R}$), and the algorithm ends.

c) If $w_1 > 1$, set $x_2 = (x_0 + x_1)/2$. We compute $w_2 = w(x_0) - w(x_1)$, and $w_3 = w(x_2) - w(x_1)$, and we come back to b), setting successively $w := w_2$ and $w := w_3$. The algorithm ends when all the computed w_i's are equal to 0 or 1. This happens because of Lemma 6.1.8.

The principal part of the cost of this algorithm is in the computation of the Sturm sequence $S(P, P')$.

6.2 RESULTANTS AND DISCRIMINANTS

The resultant is an algebraic tool allowing the elimination of a variable between two algebraic equations, that is it gives the condition (on the coefficients) under which two polynomials $P(X)$ and $Q(X)$ have a common root (real or complex).

The resultant gives then a way of defining algebraically the intersection of two algebraic surfaces, or of a curve and a surface for instance. Another elimination technique, of a more algorithmic nature, is that of Groëbner bases, also called standard bases, which we will not study here: the interested reader will find some information in [D-S-T].

In this section, we will define and give some elementary properties of resultants. The reader who wants more details can look at more involved books.

1) Resultant of two polynomials   Let $P = a_0 + a_1 X + \cdots + a_p X^p$ and $Q = b_0 + b_1 X + \cdots + b_q X^q$ be two polynomials with coefficients in a commutative ring ($\mathbf{Z}$ for instance). However, we will in general assume that the a_i's and the b_j's are independent variables, i.e., that we have $A = \mathbf{Z}[a_0, \ldots, a_p, b_0, \ldots, b_q]$.

6.2.1 Definition   *The resultant* $R(P, Q)$ *of* P *and* Q *is the element of* A *equal to the determinant of the "Sylvester matrix"* $M(P, Q)$

$$
M(P,Q) = \begin{pmatrix}
a_0 & \cdots & a_{q-1} & \cdots & a_{p-1} & a_p & & & \\
 & \ddots & & & & & \ddots & \\
 & & & a_0 & \cdots & \cdots & & a_p \\
b_0 & \cdots & \cdots & b_q & & & & \\
 & \ddots & & & & & & \\
 & & & & b_0 & \cdots & & b_q
\end{pmatrix} ;
$$

this matrix is a square matrix of order $p + q$. It has q rows of a_i's and p of b_j's.

6.2.2 Remark   If d is an integer, $\mathcal{P}_d(A)$ will denote the set of polynomials of degree $\leq d$ with coefficients in A. The degree of the polynomial P is denoted by $d(P)$. If j is an integer such that $1 \leq j \leq \inf(p, q)$, we will set Ψ_j for the linear map: $\mathcal{P}_{q-1-j} \times \mathcal{P}_{p-1-j} \longrightarrow \mathcal{P}_{p+q-1-j}$ defined by $\Psi_j(U, V) = UP + VQ$.

Considering the canonical bases of the spaces of polynomials (with the natural order on the monomials, i.e., by increasing degrees), we see that the matrix $M(P, Q)$ is the matrix of the linear map Ψ_0.

6.2.3 Proposition   a) *Assume* A *factorial (it is the case if* A *is a field, or one of the rings quoted above). Then* $R(P, Q) = 0$ *if and only if* $a_p = b_q = 0$, *or if the polynomials* P *and* Q *have a common non trivial factor in* $A[X]$.

b) *We have* $R(P,Q) = UP + VQ$ *with* $U \in A[X]$ *of degree* $\leq q - 1$, *and* $V \in A[X]$ *of degree* $\leq p - 1$.

Proof a) If $a_p = b_q = 0$, it is clear that $R(P,Q) = 0$.

If P and Q have a common factor S in $A[X]$, we may write $P = SP_1$ and $Q = SQ_1$, and the relation $Q_1 P - P_1 Q = 0$, with $d(Q_1) \leq q - 1$ and $d(P_1) \leq p - 1$. The map Ψ_0 is then not injective, which implies $R(P,Q) = 0$.

Conversely, assume $R(P,Q) = 0$, and $a_p b_q \neq 0$. As the map Ψ_0 is not injective, there exist polynomials P_1 of degree $\leq p - 1$ and Q_1 of degree $\leq q - 1$ such that $Q_1 P = P_1 Q$. This equality implies that P cannot be prime to Q (otherwise, P would divide P_1 by "Euclid's Lemma", which is absurd), and so P and Q have a common factor.

b) Let us multiply the ith column of the matrix $M(P,Q)$ by X^{i-1}. We obtain a matrix M', whose determinant is equal to $X^{\alpha} R(P,Q)$, with $\alpha = (p + q - 1)(p + q)/2$. But we may also compute the determinant of M' by adding all the columns to the first. The first column then becomes

$$P(X), XP(X), \ldots, X^{q-1}P(X), Q(X), \ldots, X^{p-1}Q(X).$$

If now we compute the determinant by development with respect to the first column, we obtain $X^{\alpha}(UP + VQ)$, with $d(U) \leq q - 1$ and $d(V) \leq p - 1$, whence $R(P,Q) = UP + VQ$.

■

We always assume $a_p b_q \neq 0$, and that we may write

$$P = a_p(X - \alpha_1)\ldots(X - \alpha_p)$$
$$Q = b_q(X - \beta_1)\ldots(X - \beta_q)$$

with α_i and β_j in a field K (it always possible in an convenient extension K of the fractions field of A). We have then

6.2.4 Proposition   *With the above notation, (and then always* $a_p b_q \neq 0$*), we have*

$$(i) \qquad R(P,Q) = a_p^q b_q^p \prod (\alpha_i - \beta_j) \quad (1 \leq i \leq p,\ 1 \leq j \leq q)$$

$$(ii) \qquad R(P,Q) = a_p^q \prod_{i=1}^{p} Q(\alpha_i) = (-1)^{pq} b_q^p \prod_{j=1}^{q} P(\beta_j)$$

Proof (ii) is an easy consequence of (i).

It is enough to prove 6.2.4 when a_p, b_q, α_i and β_j are independent variables, since, if $\psi : \mathbf{Z}[a_p, \alpha_1, \ldots, \alpha_p, b_q, \beta_q, \ldots ; \beta_q] \longrightarrow A$ is a ring homomorphism, we deduce a ring homomorphism $\phi : \mathbf{Z}[a_0, \ldots, a_p, b_0, \ldots, b_q] \longrightarrow A$, (since the a_i/a_p's and the b_j/b_q's are symmetric functions of the roots), and if $P_\phi = \phi(a_0) + \cdots + \phi(a_p)X^p$ and $Q_\phi = \phi(b_0) + \cdots + \phi(b_q)X^q$, we have $R(P_\phi, Q_\phi) = \phi(R(P,Q))$, by definition of the resultant which uses only addition and multiplication on the coefficients.

Set then $A = \mathbf{Z}[a_p, b_q, \alpha_1, \ldots, \alpha_p, \beta_1, \ldots, \beta_q]$. It is clear that $R(P,Q)$ is zero if there exists i and j such that $\alpha_i = \beta_j$ (see 6.2.2: P and Q have then a common factor $X - \alpha_i$). This implies that $R(P,Q)$ is divisible by $\alpha_i - \beta_j$, and so by $\prod(\alpha_i - \beta_j)$. But $R(P,Q)$ is divisible by $a_p^q b_q^p$, since $R(P,Q)/(a_p^q b_q^p)$ is a function of the (a_i/a_p)'s and of the (b_j/b_q)'s, and so of the roots α_i and β_j.

We deduce that $R(P,Q)/(a_p^q b_q^p)$ is divisible by $\prod(\alpha_i - \beta_j)$, and that they are equal, since they have the same degree, and same coefficient $+1$ for the monomial $(a_0/a_p)^q = \prod_{i=1}^p (\alpha_i)^q$, as it is immediately seen.

∎

b) Discriminants   Let $P = a_0 + a_1 X + \cdots + a_p X^p \in A[X]$, A being equal either to $\mathbf{Z}[a_0, \ldots, a_p]$, or to a factorial ring which we do not specify precisely.

Let $P' = a_1 + \cdots + p a_p X^{p-1}$ be the derivative polynomial of P. It is then clear from the definition that the resultant $R(P, P')$ is divisible by a_p.

6.2.5 Definition   *The discriminant $D(P)$ is defined as*

$$D(P) = \begin{cases} 0 & \text{if } a_p = 0 \\ (1/a_p)R(P, P') & \text{if } a_p \neq 0. \end{cases}$$

6.2.6 Examples
1) Set $P = aX^2 + bX + c$, $a \neq 0$. Then

$$R(P, P') = \begin{vmatrix} c & b & a \\ b & 2a & 0 \\ 0 & b & 2a \end{vmatrix} = a(4ac - b^2)$$

and then $D(P) = 4ac - b^2$.

2) Set $P = X^3 + pX + q$. It is then easy to verify that $D(P) = 4p^3 - 27q^2$.

6.2.7 Proposition   Let α_i $(1 \le i \le p)$ *be the roots of P in a field containing A. Then $D(P) = (a_p)^{2p-2} \prod_{i \ne j}(\alpha_i - \alpha_j)$.*

Proof   It is an immediate consequence of Proposition 6.2.4: we have $R(P, P') = (a_p)^{p-1} \prod_{i=1}^{p} P'(\alpha_i)$ by 6.2.4, whence the expression of Proposition 6.2.7, evaluating $P'(X)$, and replacing X by α_i.

∎

6.2.8 Remark   It is possible to generalize the resultant of two polynomials, with the notion of sub-resultants, which are determinants extracted from the Sylvester matrix $M(P, Q)$, and whose vanishing expresses that P and Q have exactly r common roots (in an algebraic closure of the fraction field of A, see for instance [B-R]).

6.3 NOTIONS OF SEMI-ALGEBRAIC SETS

6.3.1 Definition   *A set $X \subset \mathbf{R}^n$ is algebraic, if it exists polynomials $P_1, \ldots, P_k$ in $\mathbf{R}[X_1, \ldots, X_n]$ such that $X = \{x \in \mathbf{R}^n | P_i(x) = 0 \; (1 \le i \le k)\}$.*

6.3.2 Remarks

a) Any real algebraic set is the set of zeros of a unique polynomial P, namely $P = \sum_{i=1}^{k} P_i^2$.

b) The algebraic subsets of $\mathbf{R}$ are the finite sets.

c) Examples of real algebraic sets are the real algebraic curves, the real algebraic surfaces, etc ... On the contrary, a connected component of an algebraic set is not in general an algebraic set (see for instance the hyperbola $XY - 1 = 0$), and the projection of an algebraic set is not necessarily algebraic. For instance, the unit circle in $\mathbf{R}^2$ projects on the x coordinates onto the interval $[-1, 1]$, which is not algebraic (see b) above).

This remark motivates (among other things) Definition 6.3.3 below.

d) Any real algebraic set has a finite number of connected components.

More precisely, it can be proved (see [B-R]), that if $A \subset \mathbf{R}^n$ is the set of zeros of $P \in \mathbf{R}[X_1, \ldots, X_n]$, the degree of P being d, then the number of connected components of A is lower or equal to d^n.

6.3.3 Definition   *A set $A \subset \mathbf{R}^n$ is semi-algebraic if it has a presentation of the form*

$$A = \bigcup_{i=1}^{s} \bigcap_{j=1}^{r_i} \{x \in \mathbf{R}^n \; ; P_{i,j}(x) \; s_{ij} \; 0\},$$

where, for each $i = 1, \ldots, s$ and $j = 1, \ldots, r_i$:
   a) $s_{ij} \in \{>, =, <\}$;
   b) $P_{i,j} \in \mathbf{R}[X_1, \ldots, X_n]$.

In other words, the set of semi-algebraic sets of $\mathbf{R}^n$ is the smallest family of subsets of $\mathbf{R}^n$ such that:
   1) it contains the sets of the form

$$\{x \in \mathbf{R}^n; \; P(x) \geq 0\}, \quad P(X) \in \mathbf{R}[X].$$

   2) It is closed for the elementary set operations (finite unions, finite intersections, and complementarity).

6.3.4 Remarks and examples

   1) Any real algebraic subset of $\mathbf{R}^n$ is semi-algebraic.

   2) The semi-algebraic subsets of $\mathbf{R}$ are the finite unions of intervals (and points).

   3) The following assertions can be proved (see [B-R]):
   a) Any connected component of a semi-algebraic set is semi-algebraic.
   b) The interior and the closure of a semi-algebraic set are semi-algebraic.
   c) The projection of a semi-algebraic by a linear projection

$$\mathbf{R}^n \longrightarrow \mathbf{R}^{n-k}$$

is a semi-algebraic set.

   4) Any polyhedron is a semi-algebraic set.

   5) Any semi-algebraic set has a finite number of connected components which are semi-algebraic sets; to see that, we can use for instance 6.3.2, d), and the fact that any semi-algebraic set is the projection of an algebraic set.

   6) The fact that any polyhedron is semi-algebraic implies that we may approximate any compact set of $\mathbf{R}^3$ (for instance) by a semi-algebraic set:

the semi-algebraic sets let us model forms in $\mathbf{R}^3$. They have moreover some remarkable properties of stability (by projection, closure, etc.), and are in principle well adapted for the techniques of formal computation, with the methods sketched in this chapter.

They are much studied these days in theoretical robotics, but not yet much in CAD. However, they probably will be more and more used in CAD, graphics, etc.

Plates

1   Control polygon of a bicubic B-spline tensor patch.

2   The same B-spline surface in a "realistic fashion".

3   The non-uniform cubic B-spline basis elements that are used for the modelization of a "settee". Above, B-spline curve (with its control polygon) making a plane section of the settee. Note the use of the multiple nodes 3 and 6.

4   Control polyhedra of the settee.

5   The iso-parametric net of the settee. The red curve is the one of Figure 3.

6   The "realistic" settee.

7 and **8**   A ski shoe modelized with bi-cubic B-spline tensor products.

This set of plates is due to the courtesy of Dassault Systems Company.

CATIA is a registered trademark of Dassault Systems Company.

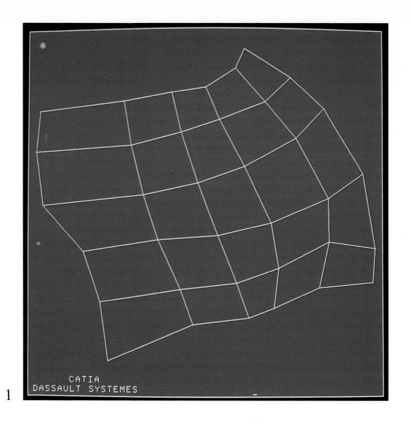

1

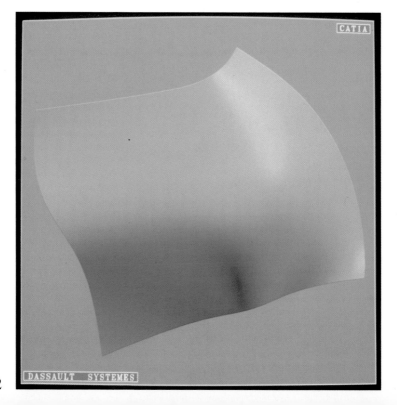

2

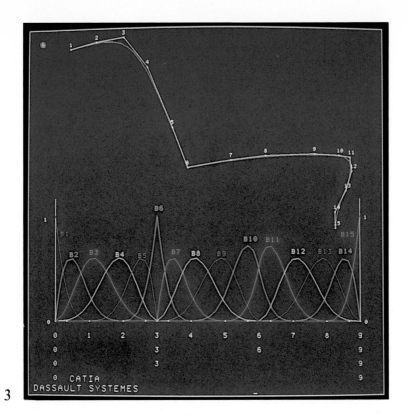

3

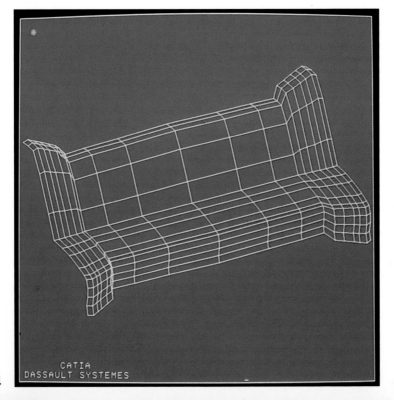

4

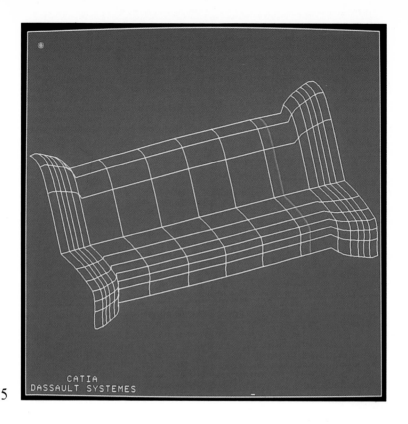

5

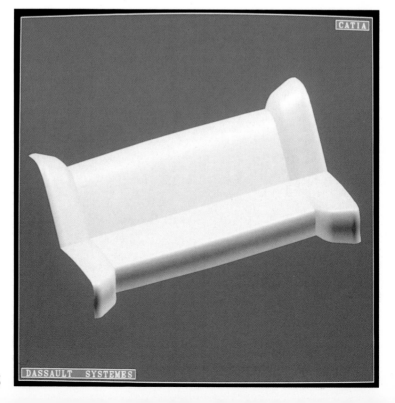

6

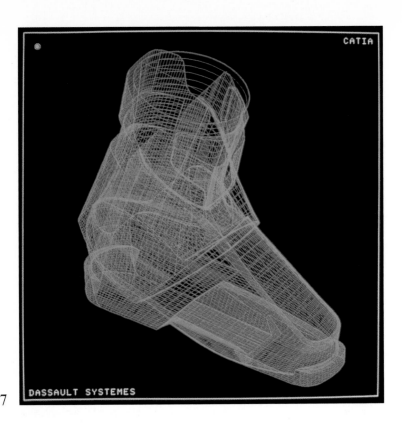

7

8

Bibliography

[B] **Becker T.J.**: Tetrahedral Mesh Generation for the Calculation of Flows around Complex Configurations. Second Nobeyama workshop of Fluid Dynamics an Supercomputer (1987).

[Ba] **Barsky B.**: Computer Graphics and Geometric Modeling using Beta-splines, Springer-Verlag (1988).

[B-R] **Benedetti R., Risler J-J.**: Real Algebraic and Semi-algebraic Sets, Hermann, Paris, (1990).

[DB] **De Boor C.**: A practical guide to Splines, Springer-Verlag, New-York (1978).

[Ci] **Ciarlet P.G.**: Introduction à l'analyse numérique matricielle et à l'optimisation, Masson, Paris, (1982).[English translation: Introduction to Numerical Linear Algebra and Optimisation, Cambridge University Press (1989)].

[DB-H] **De Boor C., Höllig K.**: B-splines without divided differences, in Geometric Modeling, edited by Farin G., SIAM (1987), pp 21-27.

[D-M] **Dahmen W., Micchelli Ch. A.**: Recent progress in Multivariate Splines, Approx. Theory IV, Edited by Chui, Schumaker, Ward, Academic press, New-York (1983), pp 27-121.

[D-S-T] **Davenport J., Siret Y., Tournier E.**: Calcul Formel, Masson, Paris (1987).

[Du] **Du W.H.**: Etude sur la représentation de surfaces complexes, Thèse, ENST Paris, groupe image (1988).

CORLET, Imprimeur, S.A.
14110 Condé-sur-Noireau
N° d'Imprimeur : 3652
Dépôt légal : décembre 1991